CONSTRUCTION
MANAGEMENT 101

CONSTRUCTION MANAGEMENT 101

A Narrative & Practical Guide to Bringing New
Production On-line On Time and On Budget

DAVID GLASS

CONSTRUCTION MANAGEMENT 101
A NARRATIVE & PRACTICAL GUIDE TO BRINGING NEW PRODUCTION ON-LINE ON TIME AND ON BUDGET

iUniverse books may be ordered through booksellers or by contacting:

iUniverse
1663 Liberty Drive
Bloomington, IN 47403
www.iuniverse.com
1-800-Authors (1-800-288-4677)

Because of the dynamic nature of the Internet, any web addresses or links contained in this book may have changed since publication and may no longer be valid. The views expressed in this work are solely those of the author and do not necessarily reflect the views of the publisher, and the publisher hereby disclaims any responsibility for them.

Any people depicted in stock imagery provided by Thinkstock are models, and such images are being used for illustrative purposes only.
Certain stock imagery © Thinkstock.

ISBN: 978-1-5320-0486-5 (sc)
ISBN: 978-1-5320-0485-8 (e)

Library of Congress Control Number: 2016915248

Print information available on the last page.

iUniverse rev. date: 10/19/2016

For My Wife Victoria

And My Children
Richard Alan, Isabelle Elena, and Louisa May
Who Never Knew What Their Dad Did for a Living

And To the Memory of Dr. Carl Wiseheart Sherman
Who Taught What it is Like to Need and Lend a Hand.

Note to title page photo: When completed in 1977, the New River Gorge Bridge near Fayetteville, West Virginia was the longest single span arch bridge in the world. At 3030 feet long, it cut the travel time across the gorge from 45 minutes to 45 seconds

TABLE OF CONTENTS

FOREWORD

Why I wrote this book and think you should read it.

I wrote this book because after spending 30 years of professional life building fast track projects with a wide variety of designers, engineers, subcontractors and owners, I gained a unique set of skills to keep the big picture on time and budget. After 10 years of experience I could predict when, where, how, and why things could go wrong and assert the kind of leadership to keep things right. Every now and then there were jobs that were bad from the start and I could only minimize the damage. Sometimes construction managers are hired after the budget and schedule has been agreed to by unqualified actors and the initial expectations are beyond any form of reality. In other situations, when there has been sufficient preconstruction planning, even complicated projects can get done with ease. This book teaches how to avoid the former and enjoy the latter; that time is indeed, money, and wasted days are wasted dollars. At Penn State I was taught how to "do the math" and at Carnegie Mellon I was taught the value of strategic thinking and lifelong learning. Now that I have the chance to reflect, and the leisure to write, I will endeavor to pass what I have learned to the next generation of engineers who build production, and the owners of capital who buy it.

PREFACE

We define construction as "to build" and a "project" "as any job that has a short term goal, limited funds, and initially an amorphous scope of work to get it done." Upper management supplies a budget and a completion on date. Whether or not either is realistic, there has been a handshake in the boardroom or on the 19th hole and we are charged with building the road to get there on time without running out of gas.

There are as many kinds and types of construction and projects as the people that put the jobs together. While the techniques and methodologies discussed herein can be useful for builders of residential and commercial big box construction projects, or IT projects with a well-defined project life cycle, the focus of this book will be on the end to end management of one-off, high stakes projects within a manufacturing or production background. The project may be initiated to capture short term profits from early adopters in a Schumpeterian market, where you make money and get out, or longer term, less risky ventures where long term growth is predicted and profits will derive from sales growth and decreasing manufacturing costs. In any case, the focus will be on getting the non-routine job done fast and with hereunto unforeseen efficiency in terms of cost and control.

The job can be big enough that unforeseen subprojects emerge as the requirements reveal themselves. Some construction sub-scopes are shovel-ready before the engineering is 100% complete. We anticipate and budget for changes, because the press of time is so severe that we must get going before everything becomes clear.

Interested readers will be corporate employees who have been designated as the project manager, whether or not they have prior project experience, or having had some prior experience completed the job late, or over budget, or in legal dispute, and wonder why. We also write to the owners of capital who wish to invest in new plant and production, as a guide to how these things should get done after the funds are approved.

It doesn't matter that the project manager had anything to do with the net present value analysis justifying the project (if there was one), or the schedule which may have been based on the overoptimistic sale of future production on an unrealistic yet specific date, or any other promises made by executives on a golf course without any knowledge of just what they were getting into on the 19th hole with that new customer.

If the job is a repeat of existing process, or expansion of something more or less proven, the future may be bright. If you are building a job with next generation production equipment and unproven processes, just throw in a set of unknown actors and you may have a real tiger by the tail.

Other interested readers, sooner or later, will be those two guys in the golf cart who have succeeded handsomely because they took the time to read the following or failed miserably and are now looking for a new job, clean out of corporate mulligans.

In short, a well-known tongue-in-cheek list of project stages goes like this, more or less consistent with Dante's descent into hell:

1. Enthusiasm,
2. Disillusionment,
3. Panic and hysteria,
4. Hunt for the guilty,
5. Punishment of the innocent, and
6. Reward for the uninvolved.

And in the end, the cycle repeats *ad nauseam* because in the owner's political context, stages 4, 5 and 6 ring true. There are no lessons learned, except by the contractors who go on to the next customers and treat them to the most expensive lunch they ever ate on your own spent coin.

This book is intended to help the reader understand the day to day skills necessary to avoid the foregoing six stages, finish the job exactly when promised, at the agreed upon price. We will do this by integrating existing processes, and understanding the who, what, when, where, how, and why of everything we do. The first chapter sets the stage and defines the project requirements. For the most part we will assume that we are presented with an existing building, or completed new facility and we are faced with building a new production facility and bringing it on line and into production.

CHAPTER 1

How to spend your time.

From time to time I'll give you some examples of things to do and things to avoid, and hopefully lots of things to think about in the meanwhile. Every now and then I'll make a statement in the form of an axiom or law. Violation of the axioms will place you in a position of peril, which may or may not be escapable. I'll also speak of two types of errors. A type one error is an error having negative project consequences, but which may be remedied by additional resources within your control, such as money shifted from one budget category to another, or the consumption of *slack* – defined herein as even more precious time, which you have built into the schedule. A type two error is an error that cannot be corrected without requiring funding beyond your control, or the addition of time which you may or may not receive without the consent of a recalcitrant customer, truculent manager, or God. The unfortunate thing is that these errors may have already been made by others and you are about to get stuck with them. Beyond customers and management, we fight against time.

AXIOM 1.

NEVER AGREE TO A SCHEDULE OR BUDGET UNLESS YOU PLAYED AN INTEGRAL PART IN BUILDING BOTH, OR HAVE HAD AMPLE TIME TO VERIFY THEY ARE REALISTIC.

At the same time you need to get started, because the owners have already placed a measure of trust in you because they offered you the role of manager. The Project has a green light from the board of directors and they are waiting for the first quarterly report from your boss. You will need to make it clear (CYA) that the budget is a moving target for the first few months and the schedule can only be decided upon completion of the basic engineering at which time the scope of work will become clear enough to finalize the budget and schedule. Call the budget a ROM (rough order of magnitude.) Experience shows that from the initial concept a budget could wind up 50% above or below the final cost and a schedule might be landed plus or minus 50% of the time initially allotted. Enormous government projects might actually jump an entire order of magnitude. On the 19th hole, the owner and customer have already written this cost and date down on a bar napkin. Both have entered the first stage of the project. Overwhelmed with this enthusiasm, they have also written down these numbers with an optimistic bias as to cost and time. The first type one error *has* been committed before they knock on your door the next Monday.

The path to the second stage looms large.

The first thing you ask for is the Net Present Value Analysis of the project. I do need to write within the context that you already know what this is. If not, take a moment and reference any entry level book on finance. A set of cash flows (investment capital) going out the door and a stream of future incomes (representing the profits) from the sales generated by the investment. All of the figures are discounted by the corporate cost of capital adjusted for risk over the length of the project. Several case scenarios have been run, and management has decided pragmatically, a best case, a worst case, and the most likely case. Any case which returns a positive net present value is justified, and, the higher the NPV, the more desirable the case. Still, the investment has been discounted at the appropriate level for risk. The Net Present Value Analysis, when done pragmatically, and independent of corporate politics is the tap root of the tree we plant here, the source of the fruits we seek. We have confidence in our proceeding, but we remain at the whim of natural and market forces such as rain, famine or fickle customers. I'll not go into project finance further at this point. Just keep in mind that market forces can make some products disappear altogether, slowly or more quickly e.g.: buggy whips, steam engines, palm pilots and film cameras. Natural forces can take your project and wipe it off the face of the earth e.g.: fires, tornados, earthquakes. To manage these types of problems there are hedge funds, futures contracts and insurance companies.

At the same time, price and demand can come together in a perfect storm to make even very short term investments return a positive NPV. Remember Crocs? Blackberry? All you have to do is predict accurately when the fad ends, the price collapses, the product is eclipsed or leapfrogged. Your finance guys and market analysts hit the analytic horizon n right on the head.

All you need to worry about at this point is the project NPV which looks something like this:

$$NPV = \sum_{t=0}^{n} \frac{(\text{Benefits - Costs})_t}{(1+r)^t}$$

where:
r = discount rate
t = year
n = analytic horizon (in years)

The NPV is our starting point for justifying anything. We will, for the moment, give it a good look and file it away. The fact that someone in the organization did one to the best of their ability is a good first step. The model shown above has many variants. We may, for example prefer to put it under a quarterly or monthly microscope for management purposes. This is a matter of preference. A good construction manager, (one that builds) will develop a schedule of cash flows for project outlays on a monthly or quarterly basis, and if our NPV analysis is laid out in the same format, we can begin to replace early budgeted costs, with more refined estimated costs, followed by actual costs, as they become known or incurred. If you make the time and build a decent spreadsheet, these inputs make an excellent forecasting tool and can be done as part of a future routine clerical duty.

If a good job has been done to this point, the cash flows for the project on completion will reflect the ROM budget on the bar napkin we have been presented by the boss. Now that we have begun to accept the project from a financial point of view, we will turn attention to some basic elements of time. A subject to which an entire chapter is devoted to later on. What should be done next? Or what must be done next?

The first thing we need to do is find out how long it takes things to happen. We are not talking about an R&D project here or the development of a new pharmaceutical. For the moment we will assume that we have sufficient in house experience relatively close at hand, a sketch of a P&ID (Process and Instrumentation Drawing or Diagram) as well as a basic equipment list. Substantial completion of the equipment list is our first major project milestone.

A good equipment list will have each and every piece of process equipment from the start of the process (receiving) to the end (shipping), plus ancillary support such as maintenance shop or QA lab equipment. The equipment list is vital to get a grip on early, because it will determine the utility requirements, which may require even more utilities equipment e.g.: boiler, air compressor, switchgear, transformer, cooling tower, chiller, pumps, waste water treatment or egads! oh really? Floor space.

If you are paying attention, the second axiom is at hand.

AXIOM 2

IF THE JOB IS TO BE SUCESSFUL, YOU WILL HAVE SPENT ABOUT 65% OF YOUR EFFORT PLANNING AND LEARNING.

COROLLARY TO AXIOM 2

THE PLANNING PROCESS DOES NOT END UNTIL WE TURN THE PLANT OVER TO PRODUCTION

At this point, it will be necessary to identify some pieces of equipment typical across different types of industry to be used later for talking points as we move along. Capital Equipment (a means to production) shall not be confused with Construction Equipment (a powered construction tool) or *construction materials*: those things installed on a project that connect the equipment to the utilities; for example, power and control wiring, cable & conduit, piping, ductwork, mezzanines, stairs, platforms, other structures, etc.

Typical Utilities Equipment:

Steam Boiler and Condensate Systems
Cooling Tower
Water Chiller
Transformer
Switchgear
Motor Control Center
Air Compressor
Fans
Vacuum pumps

Blowers
Utility pumps
Dust collectors
WWTP equipment
Filter Press
Screw Press
Scrubber
Afterburners
Fire suppression equipment
HVAC

Typical Process Equipment:

Tanks
Pressure Vessels
Reactors
Hoppers
Silos
Conveyors –Bucket, Belt, Roller, Pneumatic, Monorails
Bulk solids mixing
Liquids Mixing
Elevators
Presses
Refractory Furnace
Process Ovens
Coolers / Freezers
Heat Exchangers
Build from Scratch one-off, top secret, proprietary process machinery
Grinding and Shaping Machinery
Chrushers
Bridge Crane, forklifts, AGV's, other material handling equipment
Extruders
Stretchers
Slitters
Winders
Filling Machines
Counting Machines
Auto-weigh Machines
Cutters

Looms
Robots
Painting & Plating
Washers & Dryers
Assembly Stations
Manual Assist Devices
Wrappers and Palletizers
Baggers and Fillers
Condensers
Separators
Precipitators
Dehydrators

Instrumentation and Controls:

Safety cages, interlock devices, guards etc.
Instrumentation – Measurement and Control Devices
Process Integration Software, PLCs and SCADA

QA/QC Lab and Test Equipment:

Scales, Microscopes, Gauges, Mass Spectrometers, Jigs and Holders.

Maintenance Shop Equipment:

Lathes
Milling Machines
Drills
Welding and Burning
Special Ventilation

I'll admit to having pulled this list off the top of my head based on years of working in many different plant settings. In many cases the equipment is similar; it is the sounds, smells, and people that are different in every case and keep things interesting. As we go forward, I'll stress that we are not doing residential or commercial work. The building construction, ie: floors, walls, roof, have been completed. Office requirements are understood. Doorknobs, paint, carpet and other furnishings shall be or have been detailed and selected by others. The site selection has been completed, we have the rail siding or quay if needed, the loading docks are suitable in number and the roll-up doors selected are as big as we think we need them to be. The roof is strong enough to support anything that we (or Nature) wants to put on it. We should be able to hang what we want from the ceiling. The landlord has agreed to take care of any reasonable building modifications requested, not your problem here.

Returning to our equipment list, we now discover that this is just the beginning. We now are required to specify each type of equipment. And we start by numbering each and every piece of equipment. Equipment numbers do not come out of a hat. We must think a little like a librarian at this point:

- Utilities equipment will all be prefixed with a "U"
- Process with a "P"
- Instrumentation and controls shall be "IC"
- Lab with an "L"
- Maintenance Shop will be "MS"

The foregoing is just an example. In larger settings the process area might be broken down further:

- Wet area or Dry area
- Hot or Cold
- Hazardous or Non-Hazardous
- Clean Room.

The trick is to pick a high level designation that makes sense from the beginning.

Next is the equipment number or which we assign for very important reasons which will be revealed later on. Usually, the thing to do is to pick an acronym beginning with the first two letters of the equipment name or some other scheme that can be for convenience as long as we avoid duplication. Let's say we have two milling machines in the shop. We call them MM1 and MM2. Fully called out, they are Equipment Numbers MS-MM1 and MS-MM2. The pumps may be U-PU-1 through U-PU-20 if there are 20 pumps. There might 5 robots in a singly designated process area, we call them P-RO-1 to P-RO-5. If there is a single little robot in the clean room, he will be designated CR-RO. In this case we drop the number and you ask why? The answer is: that is how we know there is exactly one of them in the Clean Room.

Suddenly, the equipment list has turned into 5 spreadsheets, with 2 columns so far (Equipment Number and Description). We begin to understand that there are lots of things to discover. A colleague of mine once aptly said at this stage in the job: "If we don't know what we don't know, we must focus on learning".

We add columns of the things we need to know, sooner or later. While perfectly complete lists are desirable and do exist, (we may be duplicating a well-documented prior project), the lists can remain living documents, subject to change, even after procurement begins.

The equipment list, rows shortened for brevity for the process area looks like this.

Process Equipment:		1	2	3	4	5	6	7	8	9
Eqt.No.	Description	Vendor	Address	POC	e-mail	Cellphone	Office	Utilities Required	Waste Stream	Lead time
P-TA-1-5	Tanks									
P-T-A-6,7										
P-PV	Pressure Vessels									
P-RE	Reactors									
P-HO-1-6	Hoppers									
P-SI-1,2	Silos									
P-SI-3,4										
P-CO-BE	Conveyors – Belt									
P-CO-RO	Conveyors – Roller									
P-BSM	Bulk solids mixing									
P-LM	Liquids Mixing									
P-EV	Elevators									
P-PR-1-5	Presses									
P-RF	Refractory Furnace									
P-PO-1,2	Process Ovens									
P-PO-3,4										
P-CO	Coolers / Freezers									
P-HE	Heat Exchangers									
P-BS-TS	Build from Scratch, top secret.									

There are a few questions to be asked and answered at this point in our development. You will note that we have expanded the logical structure for the equipment numbers one more step beyond the earlier discussion. For example, we have seven tanks on the job, tanks 1-5 are all the same (or very nearly all the same). Tanks 6 & 7 are the same as each other, but fundamentally different than 1-5. Different in size, material, or configuration. There are six hoppers, all the same. There are four silos, 1 & 2 are alike, but different than 3 & 4. Same with the four process ovens. There is exactly one of everything else.

The next question is why do we have five lists? – Utilities, Process, Shop, Lab, Instrumentation and controls. The answer is that you are the project manager, and have reached the point where you will begin to delegate, rather than keep people in the dark, start to micro manage everything and shove equipment down the throats of the end users.

The Lab List.

We give the lab list to the lab manager, because he or she will ultimately be responsible for the performance of the lab. One microscope might have a better resolution than another, and you don't even have a clue as to the resolution required - (really the last thing you need to worry about). All of the lab equipment will have parameters pertaining to quality, accuracy, resolution, repeatability, process time calibration etc. The Lab Manager is the best one to specify and select the lab equipment. The list presented earlier (as an example) will probably grow to include everything that will go into the lab. As project manager, you will remain concerned about the utilities required. The utility consumption is likely to be small compared to any part of the process, but you will still have to get them to the proper location. City water, power, compressed air, natural gas come to mind quite easily. There may be a requirement for a special vent and duct up to the roof. You will take the list and hand it over to the Lab Manager with a memo that says "please return complete by next Tuesday. On return, you will assign the equipment numbers yourself because you know how you want it done and it will only take a few minutes.

The Shop List.

The shop list is to be treated the same as the lab list. The maintenance manager and his staff will know which machines work the best. Being skilled craftsmen that take pride in their work, they will want and deserve the best tools available, just like Paganini's Violin in the hands of Regina Carter. They know which burner nozzles don't clog up and the bits and mill ends that cut the cleanest and last the longest. They know the solvents that work well without giving them a headache and the most comfortable safety glasses that don't fog up. The shop budget is relatively small compared to everything else you have going on. Let the users decide everything and they will love you forever like a kid at Christmas. Keep in mind the note on utilities and designation of equipment numbers as per the lab. Not every shop device will need an equipment number, maybe just the larger or more expensive machines.

The Instrumentation and Controls List.

This list can have a life of its own well into the project. Some items in this category will be integral to other equipment and need only limited consideration. Then again there may be parts of the process that require special design and control beyond the capability of the vendor. For example it is common to keep a robot inside a safety cage because the motion of the robot is quite robust and could cause harm or death if someone were struck during the operation. The cage can only be opened by opening a circuit which passes through an interlock switch. This switch stops the robot; now the infeed conveyor zone just before the robot needs to time out or stop until the robot comes back on line. The robot maker is more than happy to sell you the robot and cage and switch, but does not want anything to do with the signal that goes from the switch through a PLC and to the conveyor that are supplied by others. Welcome to the world of systems integration & process controls. A fairly unsophisticated process line may contain hundreds if not thousands of signals, typically for pressure, temperature, mass flow rates, electrical resistance, physical location, weights on a scale, valve position, gallons in a vessel etc. Basically there is some kind of device that can remotely measure and transmit a signal for just about anything your heart desires. Fortunately, there are essentially four types of signals that cover 99% of applications by way of electrical wire 1) Circuit Open or 2) Closed 3) Electrical Current usually 4 to 20 milliamps or 4) Voltage from 0-24 VDC. These signals go in and out of a Process Logic Controller. The controls engineer knows what to do with them. There are more applications, where signals are sent over a wi-fi, fiber-optic cable, or some other radio frequency

A good example of a control device you have out in the garage is the safety bar on your lawnmower which is a normally open switch that closes when you squeeze it and allows the spark plug to spark and the mower to run. There is no way to have your toes or fingers under the deck with the blades turning because the bar is too far away to hold down and reach the blades with your other hand or foot unless you are faster than Muhammad Ali who could turn off the light and jump into bed before the room got dark. Then again maybe you have already defeated this federally mandated safety device by jumping the switch – keeping the bar closed with ten cents worth of duct tape.

Let's return to our own process equipment list to demonstrate just how fast the signal list can grow.

You will get the idea after just a short discussion about our tanks.

We have seven tanks containing some hazardous materials. Usually, at a minimum there will be four alarms in some form of a switch internal to the tank, even for manual operation. The alarm might be a buzzer or light.

Low Level Normally Open

(Tank level drops to 5%, switch closes, alarm sounds, Time to put more in the tank.) *The little fuel light in your car comes on.*

Critical Low Level Normally Open

(Tank level has dropped further to 2%.Switch closes, alarm sounds with more urgency. Time to put more in the tank or you will run out and upset the process) *There is not one like this in your car and you run out of gas, having ignored the first alarm a little too long.*

High Level Normally Open (This can be the same switch just installed upside down)

(Tank level rises, closing the switch at 95%, alarm sounds, better stop filling the tank.) *In the automotive example, this is the federally mandated kick-off switch located in the automatic nozzle you hold in your hand.*

Critical High Level Normally Open

(Tank has risen to 99% closing the switch, alarm sounds with urgency, a spill is imminent unless you stop filling.) *In the automotive example, the tank is topped off. If we want more gas we have to put it in another container or spill it on the ground, a felony in most jurisdictions.*

Seems funny, but the tongue in cheek example of the gas pump and the lawn mower illustrate how easy it can be to defeat an interlock, and how determined someone may be to perpetrate the act. I have a real life example of a guy who jumpered a robot fence interlock with a paperclip and was lucky to make it to the hospital. This brings us to a third axiom.

AXIOM 3.

ALARMS AND INTERLOCK DEVICES ARE THERE FOR
YOUR OWN PROTECTION AND THOSE YOU LOVE. YOU
CAN SNOOZE YOUR SMART PHONE ALL YOU WANT,
BUT NEVER, EVER, IGNORE, DISABLE OR DEFEAT
ANY OTHER ALARM OR INTERLOCK DEVICE.

Back to the tanks, we proceed to describe the attributes of a rudimentary automated system. In the most simple and manual case we have 7 tanks with 4 signals each. 28 total. Now the automation process begins and in addition the audible alarms sent from the switches the signals also go to a remote operations panel which could be anywhere in the world sent over the internet.

We keep the switches we have and add a few more devices and send the signals to a PLC and an operator's panel. The program has made it difficult, if not impossible to ignore the signals.

One signal to the controller generates a cascading set of signals (S1…..SX) in accordance with the PLC program.

Our tank is almost empty when we begin the following illustration. Level switches are as follows:

2% Full "Switch Closed" Input signal sent to the controller. S1
The controller enables and sends additional outputs:
Tank discharge pump is shut down to avoid running in a dry condition S2
Tank discharge control valve is closed S3 (The valve actuator may be electrically or pneumatically driven by way of a solenoid valve)

Affected downstream process is shut down e.g.: imagine five more signals to devices asking for whatever is in the tank….S4-S8

The discharge pump reports stopped condition. S9
The discharge control valve reports that it is closed. S10
The downstream process reports that it has shut off. S11-S15

So the 2% switch signal has initiated 14 additional I/O signals.

5% Full – audible or visual alarm only alerts the operator to potential problem or advance notice of automatic shutdown at 2%. Hardly ever heard. 2 signals S16, S17, one to the controller and another to the alarm device.

(17 Total)

10% Full – Lower Limit of Normal Operation
Switch Closed Input signal sent to the controller S18
The controller asks for some conditions all the way across the plant S19:
Other infeed conditions satisfied: delivery truck grounded S20, interlock on delivery hose indicates connected condition S22
PLC says thanks, then signals:
Infeed valve enabled S24
Infeed pump enabled S25
Infeed flow meter reset = zero, S26
All fill conditions are enabled, operator pushes the button that is labeled RUN PUMP.

Depending on how the program is written if we are wet above 10% S1-S17 can be automatically or manually reset and the discharge pump and the downstream process can re-start while the tank continues to fill.

From 10% to 90% System operates in full auto.

Let's just jump to a conclusion in this illustration.

Our three devices above, in this case, 3 simple $5.00 float switches have spawned about 26 Input/Output- henceforth I/O signals by way of the PLC. Let's make the math easy and declare that each device in our tank farm will generate +/- eight signals (7) tanks, x (6) devices each x (8) signals = 336 I/O signals.

Now we add some other issues, perhaps pH control, temperature control, pressure control, agitator motor, door position sensor…5 more devices on each tank, 8 signals, still 7 tanks = 280 more I/O signals. There are now 646 I/O signals hungry for attention like a puppy. And the signals so far are only coming from the small tank farm subroutine of the plant automation program.

Luckily,

A process engineer with PLC programming skills as well as experience in specifying control devices has just joined your team in the nick of time. 23 years old and came out of thin air.

Photo – typical automated tank farm, some utilities piping upper right, process piping over the tanks. Pipefitter working on process pump lower left.

We have taken a step towards hiring an engineering firm or drawing from in house resources if we have them.

ASIDE

My Polish/Russian grandmother used to try and keep us in the house by telling me and my sisters that there were packs of wild dogs running outside. It's true. And metaphorically there will be many of them at the factory gates trying to extract money. These beasts are called engineers, vendors, and subcontractors. A rabid She-Wolf leads the pack, chasing you straight into Dante's second ring of hell or the third phase of our project.

END OF ASIDE

If we need to sub out the engineering, we will turn to an engineering firm that has experience in our field, is close to our office, has good references, demonstrable installations and are not just another pack of dogs, or worse best friends with those guys in the golf cart. See the chapter on Procurement.

The signal list and device list begin to grow from our equipment list. As noted in the simple example above the number of devices can grow fairly quickly, but in this case we had 42 float switches that were all the same. The process engineer will have the necessary skills to select all devices. The 42 float switches could be replaced by (7) more expensive, one-per-tank capacitance probes for example.

The last thing to mention is that much of the equipment will come with its own devices and internal controls. For example the process ovens will likely have their own thermostat, ramp up and cool down controls and come with a relay panel that allows operation from a remote location by way of a push button or PLC.

AXIOM 4

WHATEVER THE SOURCE, EACH DEVICE SHALL HAVE A UNIQUE NUMBER STAMPED ON THE DEVICE OR A LEGIBLE TAG SUITABLY AFFIXED. THE DEVICE NUMBER SHALL BE CROSS REFERENCED ON ALL DRAWINGS, EQUIPMENT NUMBERS, PLANT AREAS, PURCHASE ORDERS, VENDOR DOCUMENTS AND PLC PROGRAMS.

AXIOM 5

AS A *MANAGER*, ALL YOU NEED TO KNOW ABOUT INSTRUMENTATION AND CONTROL DEVICES IS THAT THEY ARE CONSTANTLY IN PERFECT 2 WAY COMMUNICATION, SPEAKING A COMMON LANGUAGE, AND ARE PERFECTLY UNDERSTOOD BY THE PLC WHICH READS AND SENDS ALL THE MAIL EVERY OTHER MILLISECOND.

If you are going to be a good manager you might want to know a little more. There are established national standards for labeling and calling out instruments which can be found here:

https://instrumentacionhuertas.files.wordpress.com/2013/07/s_51.pdf

And that is all you, the construction manager need to know about control devices.

We return to our equipment list:

Process Equipment:		1	2	3	4	5	6	7	8	9
Eqt.No.	Description	Vendor	Address	POC	e-mail	Cellphone	Office	Utilities Required	Waste Stream	Lead time
P-TA-1-5	Tanks									
P-T-A-6,7										
P-PV	Pressure Vessels									
P-RE	Reactors									
P-HO-1-6	Hoppers									
P-SI-1,2	Silos									
P-SI-3,4										
P-CO-BE	Conveyors – Belt									
P-CO-RO	Conveyors – Roller									
P-BSM	Bulk solids mixing									
P-LM	Liquids Mixing									
P-EV	Elevators									
P-PR-1-5	Presses									
P-RF	Refractory Furnace									
P-PO-1,2	Process Ovens									
P-PO-3,4										
P-CO	Coolers / Freezers									
P-HE	Heat Exchangers									
P-BS-TS	Build from Scratch, top secret.									

At this point we have had enough internal meetings and/or have found enough vendors to have an idea about our equipment SPECIFICATIONS. What, *specifically* do we require of each item? In the example of our tanks, a complete (yet simple) specification can look like this:

In House Spec, derived from our own knowledge:

Material – Carbon Steel ASTM A-36
Coating – None
Configuration – Vertical Mount on For Legs, 45^0 cone bottom 2" 150 # ANSI
           flange discharge
Side Nozzles – (6) x ½" NPTF 2%, 5%, 10%, 90%, 95%, 98% (Float Switches)
          (2) x 1" NPTF 2%, 98% (sight glass)
          (1) x 12" Flange (Cleanout)
          (2) x 2" NPTF (immersion heater, pH probe)
Top Nozzles - (1) 24" Flange (manway)
          (1) 2" NPTF (vent)
          (1) 6" Flange (agitator)

(1) 6" Flange (emergency vent)
(1) 2" NPTF (inlet)
Quantity (5) 2000 gal, (2) 1000 gal Total 7
Diameter 48" ID
Insulation - 2" Spray On R-6 Aluminum Jacket
Service conditions 40-100 Fahrenheit, Atmospheric Pressure
Install Vendor Design Lifting Lugs, 2 places on top

Name Plate: P-TA 1-5 P-TA 6,7 Use Brass Plate 4" Font, ship loose.
Service: Tower Water 8.3 lbs/gal
Submit Shop Drawing for Approval, Current Version Adobe *pdf file.
Ship: Freight Paid, destination within Allegheny County PA.

END OF SPEC

In the case of the tanks, we have prepared the specification with in-house talent, because we have prior experience and know just what we want. We locate prospective manufacturers on the Thomas Register or Google and in this case the manufacturer is also the vendor. We are just about ready to prepare the request for quotation (RFQ) for the tanks. With a little imagination, our in-house engineers can specify the reactors, hoppers and silos. Not so with some of the other more complicated equipment. In more complicated cases, we transfer the design risk to the vendor by writing a performance specification.

The Performance Specification.

Let's construct an example supposing we are working with a garden variety glass plant. A batch mix of silica sand and other raw materials enter a furnace and are melted to an appropriate temperature and chemistry and then sent to a variety of processes, pressed into any shape you want from ashtrays (really?) to cooking bowls and dishes, beer bottles and some heavier applications such as cathode ray tubes (OK not anymore). You get the picture. We have a certain amount of breakage and scrap at a number of stations along the process. It has been going into hoppers or dumpsters for years, picked up with forklifts and carried back across the plant to a special crusher and when reduced to a common acceptable size it is reworked back into the batch mix along with the silica and other in-feeds. The broken glass (called cullet) comes with headaches. The dust is abrasive, eats rotary bearings for lunch and the small

17

shards get stuck in shoes. The larger heavier pieces erode the steel bins. The cullet also comes at us too hot to handle, creates cuts, abrasions, and silicosis. We hate to deal with it. The union and our insurance company want no more physical labor having anything to do with cullet.

The fundamental problem is to move A, B, and C streams of cullet, from five locations, each so many so many tons per hour back to the batch-house, then elevate the stream so many feet before it gets dumped. The problem is easy to see and we have been dealing with it for years. It takes five people on 3 shifts costing $50,000 per year plus the cost of at least one forklift getting all torn up each year just to move the cullet. The five bins each last one year.

The cost of dealing with the cullet is therefore:

Labor 5x3x50,000= $750,000
Forklift $25,000
Five Bins at $5,000 $25,000

$800,000 per year. Now we will look at a five year horizon, Glass has been around for a long time and continues in demand (except for those pesky cathode ray tubes which disappeared from the market in a New York Minute) so we will choose a discount rate of 10%.

The net present value of the cullet operation is as before:

$$NPV = \sum_{t=0}^{n} \frac{(\text{Benefits} - \text{Costs})_t}{(1 + r)^t}$$

where:
r = discount rate
t = year
n = analytic horizon (in years)

There are some quantifiable benefits to recycling the cullet, because it takes less batch mix for a pound of output and the overall cost of energy is reduced a little because melting down the cullet takes less energy than melting down the batch alone. For now let's assume the (benefits-costs), are reasonably inflated by 2% annually.

Net Present Value:

Year 1: $-800,000/(1.1)^0 = -800,000$
Year 2: $-800,000(1.02)/(1.1)^1 = -741,818$
Year 3: $-800,000(1.02)(1.02)/(1.1)(1.1) =- 687,867$
Year 4: -637,834
Year 5: -591,452

Total: - $3,458,971

This is the present value of the cost of the cullet reclaim over the next five years, and about what we can justify as an initial investment in today's dollars to automate the cullet return. Operating Labor is way down, the energy benefits of the recycle will remain, the bins are gone, etc. For now, let's just call the benefits equal to the much lower operating costs on a constant basis and be done with it.

We have an automation budget of $3.5 million.

The basic idea of the performance spec is to transfer the design risk to a vendor with the appropriate experience and know-how to tackle a difficult, new or unknown part of the process. Good ones can cost money, but there is competition for most things.

More on qualifying and managing vendors later.

We write performance spec that looks like this

Cullet Conveyor System:

Move 2 Streams of "A" cullet at 100-1200 Fahrenheit from manufacturing floor locations J5 and J6 to the top of the batch house at elevation 200' at column A5. 3 tons per hour, specific gravity. 2.5, mohs hardness 6.0.

Move 2 Streams of "B" cullet (same)

Move one Stream of "C" cullet (same)

Uptime 99.9% 24/7 Except for scheduled outages.
Provide dust collector at inlet and outlet and intermediate points as necessary.
End to end process loss shall be not more than __ percent of throughput.

Power available is 480 VAC 60 Hz.

Available pathways are as indicated

Provide Concept, 50% 90% and shop drawings for approval

Use equipment number P-CO – XXX, XXX. Affix name plate on each section, or not more than every 50 feet of length.

Instrumentation and control devices shall be Endress-Houser or as approved. Devices shall be tagged.

Device numbers shall be issued by Owner.

Vendor shall include in his quote, any and all spare parts, and skilled labor for maintenance for the next five years.

Cost of installation shall be by others. Vendor to include 80 hours of supervisory time during scheduled installation.

Vendor shall deliver system not more than 18 months after PO Date

Vendor shall provide a schedule of review meetings and fabrication progress meetings.

Vendor shall accommodate any unscheduled shop inspections on a six hours notice.

Vendor shall be provided as many five gallon pails of sample cullet as desired, to be picked up on 24 hours notice at (location).

Vendor shall be solely responsible to verify all physical, chemical, and thermal qualities of the cullet. Vendor shall verify all plant dimensions, as well as any and all dimensions on any drawing.

If the conveyor is to be covered, provide access doors every 20 feet.

Shipping:…

Commercial Terms 25%, Down 25%, on final design approval, 25% upon receipt, 25% after continuous running for six months. Or as otherwise mutually agreed.

END of SPEC

SHORT ASIDE ON INDUSTRIAL ROBOTICS

There will probably be another performance spec for the robots. You will provide a detailed process flow scheme and rate as well as samples of what the robot needs to handle. You let the Yaskawa or Fanuc guy pick the right robot off the shelf, strong, fast and lanky enough, with the right number of degrees of freedom. Each application will require a specific "hand" or eye or nose, (AKA end of arm tooling) that will do what you want –paint, weld, grip, mangle, bend, suck, blow, lift, place, load, unload, strike, fold, blast, burn, stack, poke, stab, inflate, stretch, wrap, kick, catch, throw and on and on. Robots are cool. Robots will work 24/7/365 without complaining. They work when it is hot, when it is cold, or raining, They are dumber than a bag of hammers. They only do what they are told. They are fast thinkers. They always show up for work and don't bitch about being bolted to the floor. They can throw up three pointers faster than Stephen Curry. The robot will have its own controller, and will receive signals from all kinds of devices and PLC's that tell it what to do and when to do it. Unlike old dogs, you can teach old robots new tricks.

Photograph Courtesy Carnegie Science Center, Pittsburgh. Used with Permission.

In the '90s I served as a construction manager for an enormous joint venture between Sony and Corning. Together, in a very short period of time they spent about $250 million to build a plant to manufacture glass panels and funnels for cathode ray tubes for the big WEGA televisions and Trinitron monitors for PC's. The market for PC's was explosive. The NPV for the expenditure was positive after 2 years, based on any estimate. The output for the plant was 16 per minute, running on two lines - 960 per hour 24/7/365.

8.4 million units per year. Better than printing money

Then the flat screens took over. In four years the market for CRT's disappeared along with the high tech fully automated production process and the plant has stood idle ever since. Years later a guy on the scene told me the only thing that did not go to scrap were the robots who were still waiting to work another day.

If they got the project horizon just right and nailed the discount rate the Joint Venture probably returned millions.

END OF ASIDE.

We have two more broad categories of equipment to deal with. The next one is the more difficult to manage, but can be lots of fun for the right engineer.

THE TOP SECRET, ONE-OFF, BUILD TO DRAWING FROM SCRATCH. THERE IS NO SPECIFICATION.

Imagine a machine like the DeLorean in back to the future - or more realistically something like the Lunar Excursion Module from the Apollo Program. A critical subcomponent in a much larger project. The machine was last built 20 years ago, but now we need to build another one, more or less the same. We have the drawings, but the subcontractors, vendors and any other project knowledge has gone unrecorded or has been lost. Maybe our machine at hand is not quite as complex as an LEM, but by the time we get the call for help, some type two errors have been made. After an intense search for a qualified vendors, and some executive wining and dining, the drawings, (500) of them are issued to 5 companies, so-called makers of "custom equipment". The drawings describe 5 machines, each one similar, but each one little different from the others. Embedded within the drawings are about 1000 purchased parts, and all of the hardware needed to put everything together.

The drawings describe a wide variety of machined parts, shafts, holders, gears, base plates, etc. all which will come together to form 5 real machines. The executives passed the drawings to the procurement department and said they wanted the request for quotation to go out to their new golf buddies in the next 24 hours with a lump-sum price for each machine to be returned within two weeks, and a delivery six months later. A hodgepodge of drawings and a single page letter requesting the bid is thrown together and let to bid. No one even looked at the drawings before they went out.

Top management from **Bidder #1** calls back after three days and say they would love to work on this job and provide a price, but need at least one month to do a responsible estimate and come up with a solid plan to meet the schedule. They are very disappointed when you tell them you cannot extend the deadline.

Bidders #2 and #3 wait until the due date and then say they are sorry, but they will not be bidding due to the complexity of the project.

Bidders #4 and #5 return a lump-sum bid, as requested, but cannot guarantee delivery in six months. Bidder #4 quotes $5.5 million, Bidder #5 is $6 million.

Then it emerges that when these machines were originally fabricated 20 years ago, the cost for all five machines was about $1 million. Adjusted for inflation, we should be looking at a price of somewhere around $2 million. Now, today, the political hindsight from the executive suite sets the budget at $2 million. Rain or shine.

This sub-project of ours has already entered stage 3 and nothing has got done at all. What went wrong and how do we recover?

Clearly, our first axiom was violated, no idea of cost or time to completion. Axiom 2 was violated in the preparation of the RFQ ie: no planning.

We must take a moment to cut through the smoke and bullshit. We will assume that each bidder had knowledge of the others, and we assume there was no nefarious communication, cooperation, or collusion through the bidding process. Things were fair and aboveboard.

Even though the bidding was a waste of time, and will not result in a PO, let's look at the bid results and see if we can learn anything.

Bidders #2 and #3 were dreamers from the start, probably small shops with one or two bread and butter customers. They were at 75% capacity before your job came in the door. They might be fully capable of making a few parts, here and there, or could have been quality subcontractor to the successful prime, but just did not have the capacity to respond. Still, the fact that they did not respond with anything until the due date indicates that they probably looked through the drawings, got lost and then frustrated by the complexity, and returned to the comfortable bosom of their bread and butter business. Their annual revenues were $1.5 million and you were asking them to do more new business than that in six months. In the end, they lacked enough horsepower and desire, and finally did you a favor by not bidding.

Bidders #4 and #5 were very large and capable manufacturers. Used to running with the big dogs they also come with very big balls. The round of golf they played with the boss was at Augusta National. Your job, relative to other things they have going on is relatively small by comparison. They put 10 well paid estimators on the job immediately and worked it on overtime. Still, when the bid was due they weren't done, so they measured their progress and the costs they had time to assemble. Then they did a rough order of magnitude estimate on what they thought remained and adjusted it for risk. While they had ample resources to get the job done on time, there was another quote for a more important project under evaluation which would have to be given priority over your job, hence they hedged a little on the delivery date. Their revenues were already $50 - $ 100 million and your business was not that important in the end. They both wound up throwing pretty big numbers at your job, and might even have been inconvenienced if you gave them the work.

Bidder #1 was your winner. They had been in business a long time. They were well run with little turnover. They saw your project as an excellent opportunity, within their in-house capability and established network of subcontractors. They valued honesty. Being conservative, they immediately asked for a little more time. Someone had to parse out all of the purchased parts, send an RFQ for the electrical parts to one vendor and the pneumatics to another. Some big round parts would go to a big round subcontractor; they could do all the plate milling on their own. Weldments would go to another guy. Only having three estimators things were just going to take a little more time to sort out and put together. Their manufacturing equipment is fully depreciated, overhead is low and their golf is played on a nice enough public course down the road. When you told them there was not going to be more time, they kept working, hoping everyone else would ask the same and they could get back into the game. There were examples of similar jobs and references you could call. If they had another two weeks to work on the quote they would have returned a fair and reasonable price of $2.5 million with a 10-15% gross margin.

AXIOM 6

HONEST PEOPLE SELDOM MAKE GOOD SALESMEN

The trouble continues and the problem escalates. You ask Bidders 3 & 4 for clarifications and guarantees. Upper Management looks for more money, but can only carve it out from another budget. Two more weeks have gone by with no action. You are about to throw about 2 million dollars into the incinerator.

Then the phone rings. The owner of Bidder #1 has discovered your dilemma and wants back in the game. He says things could have been PLANNED OUT a little better. The mistake was in asking for a lump sum from a single vendor in a short amount of time. Unconsciously, a hodgepodge of discrete and different eggs were thrown haphazard into one basket and we only asked for one big bird to sit on them. We wanted to take the easy way out and centralize the effort and it backfired.

If we took the time to fully understand the drawings and the requirements, we would have approached this job from an entirely different perspective.

Having studied the drawings for two weeks, Bidder #1 (I am calling him Bob from now on)

Has identified the basic requirements:

- • Procurement of Purchased Parts:
  - o Electrical – photo-eyes, limit switches, e-stops, drive motors, servo motors

- o Pneumatics & Hydraulics, solenoid valves, air & hydraulic cylinders, tubing, hoses
- o Motion Controls / Power transmission. slides, belts, pulleys, pillow blocks, bearings, chains, cogs, gearboxes, motor couplings etc.
- Custom Machine Parts - many types and sizes. Some as shown, some opposite hand, some one-off, some repetitive and appear many times
- Structural Fabrication and Weldments
- Hardware – M1x2 to M30x100 and everything in between
- Assembly and Sub assembly of the above

Installation of all of our equipment will be discussed in subsequent chapters on construction

Bob has taken the 500 drawings, each containing one or many more of the above requirements and has started to construct a *more coherent and comprehensive* Bill of Materials (henceforth BOM). He has discovered, nay uncovered! That most all of the purchased parts on the drawings are called out somewhere by manufacturer and even part number, we just had to find and consolidate them.

People can easily get confused about BOM's, and it takes some experience to get used to them.

A BOM on a simple drawing will call out all of the materials to make what is shown on the drawing.

8x4 child's sandbox:

BOM:
(3) – 8' x 12" x 2" Pressure treated #1 Pine.
(12) – 20d Common Nail.

The drawing will show one of the boards cut in half and placed in the rectangular configuration and fixed with 3 nails driven into each of the four corners.

If you were building a whole playset, there would be another drawing for the rain cover, swings, slide and ladder, each with its own BOM. These different pieces of the playset are called sub- assemblies. If the sandbox above were part of the playset, it will have become a sub-assembly. The whole playset is called the Assembly. There may or may not be a unique BOM for the entire assembly if all we need to put it together is hardware. Then, all we might need is a hardware list that we can compare to what we got in the box like we did when we bought the shelves at IKEA.

Our case of course is a little more complicated, but really not by much. Purchased parts have been divided into three categories: electric, pneumatic & hydraulic, and motion controls. Bob will complete a BOM for everything within the three categories and get them out to bid. Over the years, industrial supply vendors and manufacturers have evolved to make and sell according to these categories. Omron is good for photo-eyes PLC's and switches, SMC pneumatics, Applied Industrial will be able to vend just about anything on the list. A walk around the web will get you acquainted with industrial suppliers. Some manufacturers sell through company owned branches, some through licensed distributors, others through vendors. Sorting everything out would be a perfect job for a young engineer as they don't get this kind of hands-on commercial experience in college. Call him or her your Project Engineer.

Further, http://www.mcmaster.com is one of several industrial equivalents to amazon where you can find almost anything, http://www.grainger.com is another that may not have as much in stock, but may have a local branch. www.mscdirect.com is another player in industrial parts. These suppliers are more associated with plant maintenance, a little more expensive but can get you most items same or next day.

Get 2 or 3 vendors lined up and make them compete. Just about any purchased part you want is on the shelf somewhere in the world and you can have it in about two weeks. With the drawings you have, it will have taken more time to build the list and identify the vendors than it will have of get everything once the requirements have been established. I hope by now you are starting to see how easy it was to waste time by just throwing the project out to our bidders 1-5

AXIOM 7

IN THE RFQ FOR THE PURCHASED PARTS YOU SHALL SPECIFY TO THE VENDOR THAT HE OR SHE SHALL TAG EACH PART WITH THE ORIGINATING DRAWING NUMBER, INCLUDE IT ON THE PACKING LIST AND ON ALL INVOICES.

That's so you know where they go when they arrive, even if the guy on the receiving dock takes off the packing list and doesn't know what to do with it for a week. Someone else shoves the pallet in a dark corner of the plant because it got in their way. More on the subject of receiving to follow.

Choose to makeup one BOM for all the purchased parts, (List of Purchased Parts) or *Choose* to make a BOM for each category (List of electric parts, list of pneumatic parts, list of motion control parts) *Do not Choose* to make up a BOM machine by machine and wind up with five that look almost alike but with slightly different quantities. The latter is another type one error.

ASIDE

The BOM for the whole job actually might have been completed as part of the drawings a long time ago but got lost along the way. Before there were personal computers, a project like this would have had a file of many different sized paper drawings, rolled up or kept on racks, and a book of lists kept separately in three ring binders. Drawings seemed to have kept their significance over the years, while associated binders were discarded. In this case we have reconstructed the consolidated BOM from the drawings. Another thing that might be missing is the actual drawing list. If it has been lost, we will never know if we have a complete set. We can make a good guess however. A sub assembly will always reference the individual drawings that make it up, and the assembly drawings will reference the sub-assemblies. Some companies do a much better job of keeping track than others. Also, there are issues when drawings got revised, but the revisions do not get properly recorded on the lists. Then you go to order a purchased part and the vendor replies that that number has been discontinued long ago. You might find the change was in number only, or maybe the material was changed from cotton to Kevlar. Maybe the part is just obsolete. The mechanical thermostat is obsolete and replaced with an electronic one requiring power. Remember the device list lives and breathes. Head's up! Learn to expect the unexpected.

Good engineers are problem solvers and when problems emerge, they will come up with solutions. Worry about the guy who threw the first bid package out the door without looking at the drawings. Or the one who complains about not having enough information to get going. There is more on the nature of drawing control and engineers to follow.

END OF ASIDE

Now we return to the custom machine parts, and I'll turn to a nice automotive example to illustrate. Imagine your rear wheel drive car from the transmission to the rear wheels. We have a big aluminum casting (Housing), a box of gears in another aluminum casting (gear-box) connected to a shaft, by way of two universal joints, to a dual right angle gearbox or transaxle (steel casting and gears), connected to two more shafts (call them axles) which connect to the wheels. Then there is another assembly called the frame which carries all the above.

Keeping it simple, from a manufacturing point of view, we have

BOM: Item Number, Quantity, Description

1. 2 Aluminum castings (Housing, Gearbox)
2. 1 Cast Steel casting (Transaxle, 2 pieces)
3. 3 shafts (Drive shaft and two axles)
4. 10 gears six forward, one reverse, (in gearbox) crown drive, plus right and left driven gears (in transaxle casing)
5. 2 steel stampings or two more aluminum castings (the wheels)
6. 1 Stamped and rolled weldment, (the frame)
7. 1 lot Assorted Hardware.
8. 2 Universal joints (purchased parts).

The thing I am driving at (egad! the Pun!) is that machine shops come in all shapes and sizes, evolve or perish at the whim of market forces, and arrive at different specialties. Some have their own foundry operations and can cast and finish aluminum or cast iron. Some are good at turning operations and will a do good job with our shafts. Some are fitted to do plate machining but can only turn up to 12". Some can only bend sheet to a ½", but can very efficiently cut anything out of sheet with a laser table. Others are heavily invested in CNC machines that can do repetitive operations. Dana Corporation is very good at stamping out frames and shafts for Ford and GM, and doing millions a month with 100% automation. The Number 2 Machine Shop at Bethlehem steel had what it took to make the turrets and barrels for the USS Missouri.

Some are near
some are far,
some can't even bend a bar!

Can they, can they, drill my plate?
How many holes? we might need eight!

Can they read the BOM?
I don't know, go ask your mom!

Compliments Ted Geisel

The best shops are booked solid for the next six months, others are hungry. Some will do the plating in house, some will sub out the paint. Most don't advertise and serve mostly the local industry, they are there, but can be a little difficult to find. None of them have the resources or structure to do everything we want on time and at a reasonable cost. Bidders 3 and 4 told you this with their bids a month ago, but did not give you the details as to why. Dana would not be interested in three shafts, but might light up at the idea of 300 or 300 thousand

ASIDE

Look at the shops as your instruments in a symphony, each playing his part. Your RFQ and documents are the music. You are the music director and Bob wants to be conductor or concertmaster.

We pause and imagine, Gershwin at his own piano, Stravinsky conducting his own Firebird Suite. Brian Wilson and his *Smile*.

END OF ASIDE

Now we have Bob and his plan. Bob wants to make the parts he is good at, and has an area available to assemble everything. Bob knows some additional shops you never heard of. Bidders 2&3, were non responsive, but Bob knows how to get some work out of them. Bidders 4&5 didn't even call back to check on the results, let alone respond to your request for a schedule guarantee. They have been summarily disqualified from further consideration. Your project engineer gets on the web and phone and gets to know at least 3 shops in each category that can help us:

Casting
Turning
Plate
Sheet
Stamping / Welding
High volume Small volume
Big Parts Small Parts

Your project engineer is *listening* to the vendors: she can tell by the intonation in the voice, or the rank of who responds, or how long it takes for a call to be returned whether or not someone is interested - or just being (oh shit, gasp!) *Minnesota Nice*. We know to avoid vendors that worry about a credit application or terms of payment on the first call before they have received the RFQ or looked at the drawings because they are the ones most likely to be undercapitalized.

Bob is going to emerge as our lead shop who will make some of the parts in house, receive, inspect, and manage the inventory coming from the others. Bob has suitable space with good lighting and an overhead crane, his machinists will be able to do the assembly. Bob is close by. His proximity has eliminated thousands of dollars in jet travel for assembly inspections.

We are now ready to build the bid documents, and write the music so to speak.

The bid form for the machined parts is almost the same as the bill of materials, only we add a column for the unit price and the extended price. Remember we have 5 machines. Lets call them Drive Stations for convenience P-DS-1 through P-DS-5 in accordance with our established nomenclature. Remember P-DS1, 2 & 5 are exactly the same, P-DS 3 & 4 are a little different. Our machined parts consolidation begins to take shape and looks more or less like this:

1	2	3	4	5	6	7	8
Line No.	**Description**	**Quantity**	**Drawing Number**	**Machine Number**	**Tag**	**Unit Price**	**Extended Price**
1	Base Plate	3	P-DS-BP 1-2-5	P-DS-1-2-5	P-DS-BP-1-2-5	$ 2,000.00	$ 6,000.00
2	Base Plate	2	P-DS-BP-3, (4OPP)	P-DS-3-4	P-DS-BP-3-4	$ 1,500.00	$ 3,000.00
3	Transmission Gear	5	P-DS-TG	P-DS-1-2-3-4-5	P-DS-TG-1-2-3-4-5	$ 500.00	$ 2,500.00
4	Transmission Gear (2)	1	P-DS-TG2	P-DS-3	P-DS-3-TG2	$ 200.00	$ 200.00
5	Transmission Gear (3)	1	P-DS-TG3	P-DS-4	P-DS-4-TG3	$ 200.00	$ 200.00
6	Main Shaft	3	P-DS-MS-1-2-5	P-DS-1-2-5	P-DS-1-2-5	$ 500.00	$ 1,500.00
7	Driven Shaft	6	P-DS-DRS-1-2-5	P-DS-1-2-5	P-DS-DRS-1-2-5	$ 400.00	$ 2,400.00
8	Main Shaft (2)	2	P-DS-MS2-3-4	P-DS-3-4	P-DS-MS-3-4	$ 500.00	$ 1,000.00
9	Driven Shaft (2)	4	P-DS-DRS	P-DS-3-4	P-DS-DRS-3-4	$ 400.00	$ 1,600.00
10	Frame	3	P-DS-FR-1-2-5	P-DS-1-2-5	P-DS-FR-1-2-5	$ 600.00	$ 1,800.00
11	Frame	2	P-DS-FR-3, (4OPP)	P-DS-3-4	P-DS-FR-3-4	$ 650.00	$ 1,300.00
11-499	Blah	Blah	Blah	Blah	Blah		
500	Name Plate	5	P-DS-NP	P-DS-1-2-3-4-5	N/A	$ 150.00	$ 750.00

We will refer to this a figure 1 for future reference.

Our BOM is rich with unseen information, so a couple of notes are necessary before we continue:

Lines one and two have just fully described a single base plate for each machine. Plates for DS 1, 2, 5 are identical in all respects and completely interchangeable. They *should* however be tagged uniquely. Why? Because it is the best way to handle a mis-fabrication or other reason for return or unexpected transshipment. We want to talk over and correspond to a specific base plate even though others may be the same.

Drawing P-DS-BP-3 describes the base plate for DS 3&4. The plate for DS 4 is an opposite hand or mirror image of DS 3. Even though we can hold the drawing up to a mirror, or flip it over and look at it through a window, we call out "Opposite hand" because it is a two dimensional representation of what actually exists in three dimensions. One might think we could just flip one plate or the other to accommodate subsequent installation, we cannot, even if the tap holes go all the way through the plate. The "down side" (on the third dimension) might have one finish, (Paint) the "up side" another (milled to spec). This goes to illustrate that even though the Drawing is the same, these tag numbers MUST be different. Later, when the tag number is translated to a call-out on a construction drawing we avoid another type one or two error of installing both plates upside down and in the wrong location – and, oh, shit! Already cast into concrete or welded to something else. We asked for the plates two weeks ahead of the machines so we would be ready to bolt them down on arrival. The simple error of improperly tagged plates incubates for two weeks before discovery and takes another to correct. Upper management only notices the rework and would never understand what just went wrong.

This sub-project just leap frogged straight to stage four, and it doesn't mean squat how good you are doing with everything else.

ASIDE

The best laid plans of mice and men often stray awry.
John Steinbeck 1937, see also Robert Burns 1785

END OF ASIDE

Column 3 tells you that a single shipment of lines 1-11 will have contain 32 individual pieces.

Lines 1-11 taken together should be sufficient to demonstrate an effective and fairly logical sequence has emerged. Other schemes can be selected for brevity and clarity. I have endeavored to provide a structure to allow for the control of multiple assemblies, on multiple machines going to multiple areas of the site.

Lines 12-499 Tell us we have a fairly long way to go, however the logic and the repetitive nature of the scheme is revealed, and the actual effort in putting this together picks up some steam.

Figure 1, lines 1-500 and columns 1-6 represent, the (machined parts) together with our purchased parts list(s) and our hardware list complete our BOM for these five machines.

We add 7&8 columns 7&8 to fig 1 and bid form for the machined parts. This gets sent to our selected shops. With the RFQ and Instructions For Bid. (IFB)

We close the discussion on the BOM with another aside on Line 500.

ASIDE

In Heaven, everything is perfect This last drawing would show the size and thickness of the plate, call-out four holes for mounting hardware and carry a Note 1: "Emboss on 16 Gauge Stainless type 304 Use Tahoma #48 Font" Note 2 - see drawings P-DS-FR-1-2-5 and P-DS-FR-3,(4OPP) for location. Our frame drawings will have four pre-drilled holes, M2x4, matching the plate spacing, and our hardware list will have exactly 20 M2x4 bolts, because they are used in those 4 places on our 5 machines and nowhere else.

In another reality we could drop this drawing altogether and get the nameplates from McMaster Carr, or Seton and stick them on anywhere at the last minute. Or just borrow a label-maker from the shop, and save the $750.00 on line 200, column 8 and the other hidden $500 for the tap holes and subsequent installation. The Federal Government would never allow this to happen, don't even suggest it.

END OF ASIDE.

Having nearly completed the purchased parts list(s) the machined parts list, the in house specifications and the performance specifications, we are one step closer to being ready for the procurement to begin. Each piece of equipment will by now have its own prospective vendors, complete with contact information. We have spec sheets and or cut sheets from the initial vendors, and without bidding anything we have begun to arrive at the utilities requirements for our job. What comes next is to add them up and determine if our existing utilities can meet the new demand. The new demand may require additional equipment, or upgrades to existing systems. This is the point where we determine if we need a new air compressor, water pumps, WWTP equipment etc.

Like the development of the BOM, utilities can be a little confusing to the inexperienced non-engineers like those guys on the 19th hole or the accountants on the second floor. Most homes have the following utilities:

Potable WATER. Maybe from the well, city supply, or cistern
Electric POWER
NATURAL GAS. Or oil heat, or neither.
Air conditioning, if separate from the heating system
SANITARY SEWER.
Telephone cable etc. can be thought of as a utility.

Your plant may have a number of utilities, some derived or reduced from another:

- City Water – Potable, comes in from the city and goes out the drain or through the process
- Tower Water – Water that is recirculated through a cooling loop by way of a pump and evaporative cooler. Evaporative losses are made up with city water.
- Chilled Water – Water cooled below 32°F mixed with an antifreeze, recirculated, may be pre-cooled by way of a heat exchanger and tower water
- Fire Water – city water through fire pumps for suppression system.
- Nano or Filtered water – Purified water for food/drink applications.
- Process Waste Water
- Sanitary Sewer
- Low Pressure Gas – Small ovens, Burners etc.
- High Pressure Gas – Bigger applications, Furnace Boiler Rotary Kiln.
- Compressed Air
- Instrument Air - Specially filtered, regulated, and lubricated form of compressed air
- Low Pressure Steam < 50 psig
- High Pressure Steam > 50psig
- Culinary Steam -. Pure enough to meet FDA standards for contacting food.
- Industrial and Cryogenic Gases - Oxygen, Argon, Carbon Dioxide, Neon, etc.
- Electric Power
- "Clean" Power

We return to our process equipment list to build the example.

Process Equipment:		1	2	3	4	5	6	7	8	9
Eqt.No.	Description	Vendor	Address	POC	e-mail	Cellphone	Office	Utilities Required	Waste Stream	Lead time
P-TA-1-5	Tanks									
P-T-A-6,7										
P-PV	Pressure Vessels									
P-RE	Reactors									
P-HO-1-6	Hoppers									
P-SI-1,2	Silos									
P-SI-3,4										
P-CO-BE	Conveyors – Belt									
P-CO-RO	Conveyors – Roller									
P-BSM	Bulk solids mixing									
P-LM	Liquids Mixing									
P-EV	Elevators									
P-PR-1-5	Presses									
P-RF	Refractory Furnace									
P-PO-1,2	Process Ovens									
P-PO-3,4										
P-CO	Coolers / Freezers									
P-HE	Heat Exchangers									
P-BS-TS	Build from Scratch, top secret.									

Two of the tanks are heated by an internal heat exchanger, consuming 40lbs/hour of low pressure steam. To make the steam we need another 5MCF/d of gas. The agitators on the tanks draw 5 horsepower- 80 kilowatts each. The pressure vessels just sit there.….The reactors require 200 gal per hour of chilled water at 20° F. Oh shit! We never needed chilled water before, we have to buy a chiller. The chiller requires so many kilowatts of power, assuming an inlet temp of, X degrees Fahrenheit, coming from the now too small tower, means we might need an (*oh shit!*) another cooling tower, which requires another 80 kilowatt motor. Or since the existing tower is near the end of its service life, maybe we buy a single tower with a capacity for our total requirement. The silos need 50 KW each, at peak.

Conveyors, Liquids Mixing, Bulk Solids Mixing Elevators, Presses, total 500 KW. Presses need Tower water for cooling. Liquid Mixing needs so many GPM of Nano water. We have to buy more capacity for our existing nano water (Reverse Osmosis) skid.

Refractory furnace and Ovens need 50MCF/dy natural gas at high pressure.

The coolers and freezers require additional power only. The heat exchangers don't need anything, they just transfer heat, according to the first law of thermodynamics.

Finally our top secret equipment with the giant parts lists only had a few 5 KW motors. And a few air cylinders.

Outside of the main power requirements, just about everything will require a small amount of Instrument air and 100 VAC power to run the instruments.

What all of the foregoing means is that our capable project engineer will work with the outside electrical and mechanical engineers to see if we need to buy additional utilities equipment as the process requirements emerge. In our example, we definitely need a new water chiller, we might need a larger air compressor. There is plenty of power out on the street, but we might need a larger main transformer from the non-performing power company and additional main breakers inside the plant. There is probably more than enough gas on the street, but our boiler might be too small. Only when the requirements of the process equipment has become known can we learn the sum of the utilities scope. Not before. These time consuming and emergent financial requirements never ever occurred to the golf buddies back in the preface to this chapter. Maybe the utilities scope of work emerges as an unbudgeted subproject. Maybe there was a line item in the budget set-aside for the utilities, double of what you will need. Maybe you are the only one paying attention to the budget at this level. If you are the only one aware of the surplus, keep the knowledge in your strategic arsenal. You might need it later. As your humble and obedient author, I encourage you to flip back and take a fresh look at our first Axiom. You will have to commit to a budget sooner or later. For now we continue to build our budget and schedule. We have just spent a month or two learning these things we did not know.

This Chapter was titled How to spend your time. What lessons were learned?

1. Uncover all of the requirements as quickly as possible. Plan the attack.
2. You must filter and un-complicate what seems complex. After all it was only other humans that built the house to begin with. Others will not trifle with your requirements unless they see immediate and substantial profit.
3. Recognize the hard work and do it *yourself.* Conceptualize the lists that need to be done. These are the elements of *your* scope of work.
4. Know the job better than anyone to the best of your ability.

5. Recognize the easy and more specialized work and delegate to others – the scope in the shop, the lab, the device list, the controls.
6. If two or three of your senses report bullshit in front of you, it probably is so.
7. Keep the door open and recognize sincerity. Guys like bob will help you.
8. Do not lament or dwell on time wasted by others. Forgive them father, they know not what they do. Move forward and keep on the attack

CHAPTER 2

Building a Head of Steam, Understanding Engineers, and Getting Organized.

As we move forward, we will proceed on the premise that we rely on paper for only a few applications. It remains that paper is easy to mark up and scan and then send electronically when dealing with RFI's (Requests for Information) and a few drawings, which contain more information than we care to digest view through a monitor. If we are in an existing plant, the next thing to emerge will be a revised plant layout drawing where we show the locations of all new equipment, utility pathways, etc. There is always a layout drawing fixed to the wall so people can stand and talk over it.

Paper is one thing; files are another. You are going to need lots of files and an efficient system so that you can talk over just about any document within seconds of getting a call from anyone.

The first one is your contact list. This is already partly done. We just cut and paste from the equipment list.

Now, add the Newcomers:

- Bob
- Project team members
- The engineers
- The architect
- The shops
- The vendors
- The inspectors
- The roll-off box people
- The rental company
- The local hardware store
- The prospective subcontractors
- The roof penetrating guy

- The port a pot guy
- Anyone else you will need to talk with more than once.

You can have most of this on your smart phone if you want. I continue to prefer an excel file, which I keep right on my desktop. If it takes more than 5 seconds to get your hand on a phone number, you are wasting time. At the time of this writing, voice recognition software is improving at a rapid pace, but for the time being the jury is still out on Siri and Cortana

Now, make a file for every vendor. These files will come to contain all of the correspondence, bids, quotes, revisions, RFQ's RFI's, IFB, invoices, etc., for each individual vendor. You will require that all drawings are sent in a pdf format, since just about everyone is using that now. Print whatever you want, but keep everything on your laptop.

Other files will emerge, correspondence with the permit office should go in its own file.

You might want to keep all of the drawings from all sources in a single drawing file with multiple sub-folders.

Maybe all of the commercial paper goes in one folder.

You already know there will be lots of lists. Make a file called lists, and keep all of the lists there.

You need a file with a map to the hospital. When the subcontractors come on site, that one gets handed out to everyone and is posted in every gang box too. Here is a screen grab from a job I did at a beverage plant a couple of years ago. 29 files, and many more sub-files emerged before the job got done. The contact list is another kind of living document.

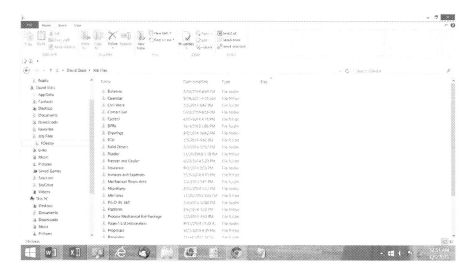

Another thing to note is that as business proceeds and equipment gets installed, some files are referenced less and less often. Put these folders in a file called "Finished Business." That way you will not have so many folders cluttering up your actual or electronic desktop when you need to get at something.

I do not presume to tell you there is one way to keep your files. Some companies will have systems in place that make you do things a certain way. There may be an "Enterprise Solution" in place that does the filing for you. Any filing system you choose however, *must* obey our next axiom:

AXIOM 8

IF THE INFORMATION IS NOT IN YOUR HEAD IT MUST BE RIGHT AT YOUR FINGERTIPS.

If you get a call from someone who wants to talk over a document, and you have to call back because you can't find it right away, that's a pretty good example of what I will call a *Creeping Error*. Maybe it takes only 20 seconds to find a document, but it happens 20 times a day, then you have to call back, and leave a message and wait again. 400 seconds a day x 5 days a week = 2000 seconds a week. You lost the chance to go home a half hour early on Friday. The business uses of text messaging are still emerging, but one way you can manage this timewaster is by sending a text: *I'd like to talk over your proposal at 10:00 please be prepared, thanks.*

Up to now, we have had the in-house power to come up with the equipment list; we knew enough to count up how many kilowatts of power are required, and how many more pounds of steam per hour are necessary. We have been able to go to the engineering handbooks and do some rudimentary heat transfer calculations. We know that one horsepower (measured at the output end of a shaft) is equal to 746 watts of electricity. We know that an electric motor has an efficiency or power factor that depends on its inherent design as well as the frequency of the alternating current fed to it. We understand the nature of these variables, but not necessarily all of the interactions between them. We can estimate the mass flow rates and heat transfer properties of our mechanical utilities; we have a good in-house process engineer, he knows our process, but the requirements associated with completing the project design require a greater diversity of engineering talent and also come with time constraints. We need several different engineers for a short period of time, and we will need to hire an engineering company to help us. Our own process engineer was educated in one of the fundamental disciplines, and will have adequate training to supervise the overall effort. He or she is not incapable of returning to the basic mathematics and physics to do everything required; there is just not enough time, and it is better to turn to specialists. We don't want a radiologist doing gynecology any more than we would want an electrical engineer sizing a sewer pipe. Doctors, lawyers, policemen and firefighters enjoy a lot of exposure in popular media: heroes, larger than life. Engineers only turn up on the history channel when things go wrong, or come from outer space and get on ancient aliens.

So let's take a moment and go into a little detail on what engineers actually do. The current 2015 definition per Merriam Webster:

engineer

noun en·gi·neer \ˌen-jə-ˈnir\

1): a person who has scientific training and who designs and builds complicated products, machines, systems, or structures: a person who specializes in a branch of engineering
2): a person who runs or is in charge of an engine in an airplane, a ship, etc.
3): a person who runs a train

2 & 3 are in the process of being replaced by computers, which is probably a good thing. We shall spend some time exploring the first definition.

I have grown to dislike the words complicated and complex. Good engineers *solve* problems by breaking them down to their more simple *elements,* like nuts and bolts or natural laws. Politicians and pundits take these seemingly simple things (elements of our faith and belief, or how we stand on the deployment of scarce resources) and make them appear so complex that we begin to face new kinds of problems. Complexity is created by man. For centuries, societies have believed in four or five elements: earth, wind, fire, water… later "the ether." Not far off the mark really, but, at the very fundamental limit of things, all physical interactions can be adequately described by combinations or derivatives of just three. These being mass, length, and time. These problems of mass (how strong should we make the bridge), length (how far to Mars), and time (tomorrow really?) are solved by the engineer at the elemental level.

Everyone who ever got a speeding ticket knows that speed is length/time. Length squared is area. Length cubed is volume. Mass/volume is density.

Remember Newton: *Force* (the result) = mass x acceleration or, F=mass(length/time^2), Finally F=ma

And Einstein: *Energy* (result) = mass x the square of speed = m (l/t)*(l/t), Finally E=mc^2

And, finally, James Watt determined that Mechanical *Power* (result) = (force x length)/time.

There is horsepower and there is manpower. A substantial discussion on the latter will follow.

And the reader will note that I have avoided the use of any conversion factors or systems of measure to illustrate just how far we can go with our three elements of mass, length, and time. Let the engineers worry about all that, because they created the mess to begin with. There is an imperial way to arrive at horsepower, and a metric one. There may very well be Imperial Horses at places like Buckingham Palace, but I have never seen a metric horse.

ASIDE

Rocket Science de-mystified, once and for all (of you!). The Apollo Saturn V rocket weighed about 6.2 million pounds on the pad. At rest, there is an equal and opposite force, the ground pushes back on our rocket, and it *cannot* move. When the (5) F-1 Engines were ignited, 7.5 million pounds of thrust were generated, and this equilibrium was upset.

Photo: Werner Von Braun stands before a cluster of F-1 engines. The fuel pumps developed 55,000 horsepower each.

The detached rocket *must* move. And they did, slowly at first, but note that the acceleration was increasing until burnout. Why? When the big candle was lit the force generated by the engines was a constant 7.5 million pounds, while the mass of our rocket was decreasing rapidly. It burned 28,415 pounds of fuel every second. Simply put, Newton's Law above must be satisfied at all times. The rocket *must* go faster and faster until it runs out of gas.

On the pad, the thrust to weight ratio was (7.5/6.2) = *1.2*

For the missions carrying a lunar module package burnout occurred after 165 seconds. So just before burnout of the first stage, the thrust to weight ratio was (7.5/(6.2-(.028415x165)) = *4.96!* and the velocity was 1.71 miles per second. Some ride!

END OF ASIDE

Thus, it is demonstrated that even the most complicated systems can be easily understood if reduced to our three simple elements of mass, length, and time. These are the fundamental kinds of problems all engineers deal with. Note how we demystified our special machine by parsing out all of the hardware, purchased parts and machined parts. That part of our job was broken down into its elements. People will be able to comprehend our requirement.

There are many branches of engineering, but it is generally accepted that they stem from four. Briefly, for the uninformed:

Electrical Engineers deal with the motion and behavior of electrons. They solve problems associated with power transmission and distribution from millions of volts down to bits and bytes. A computer scientist or software engineer can be thought of as cousins. The first electrical engineer was probably Benjamin Franklin, who saved millions by inventing the lightning rod. The next were Thomas Edison, Nicola Tesla and George Westinghouse.

Civil Engineers deal with problems of structure, roads, dams, bridges and tunnels, sewers and water. Architects are a kind of civil engineer that deal with human enclosures and how they can be made efficiently and pleasing. The first civil engineers were the Egyptians, Greeks and Romans. Some of their buildings still stand, and the Romans solved the problems of basic sanitation and running water. Some say the first civil engineers were some unnamed Australopithecus who might have thrown a tree across a river, creating the first bridge.

Chemical Engineers design and produce chemicals that we use every day, acids, caustics, detergents, and petrochemicals. The first chemical engineer were probably some Babylonians who left behind a recipe for soap around 2800 BC.

Mechanical Engineers deal with all types of machinery, and solve problems associated with the transfer and distribution of heat and the control of pressure. The first mechanical engineer was probably Archimedes, (287-212 BC) who figured out what made things float, and came up with the first lever and water pump.

Then we have our sub-branches and hybrids:

Mining Engineers deal with the efficient and clean extraction of useful minerals from the earth. The first mining engineer lived before records were kept, and dug for flint, or quarried the stones for the pyramids.

Petroleum Engineers study the flow of liquids through rocks, which are found by the geologist or geophysicist, and deliver the liquids to the refinery where the chemical engineer may take over. The first Petroleum engineer was Edwin Drake, who dug his hole in 1859 near Titusville Pennsylvania.

Metallurgists study the chemistry of liquid metals and turn them into more useful and ever stronger end products. The first metallurgist replaced flint with alloys of copper and tin, turning the page from the stone to the bronze age.

Material Scientists do the same thing with non-metals, such as plastics and ceramics. The first material scientist probably built the first boat that did not leak.

Aerospace Engineers design systems that allow people to fly and go into outer space. The first *successful* aerospace engineers were the Wright Brothers. The fathers of the field were dreamers - Icarus and Daedalus, while Da Vinci left behind some of the earliest drawings of record.

Da Vinci glider c. 1500   First Powered Flight, December 17, 1903

Just pause for a moment and imagine the life of Orville Wright. He and his brother solved the problem of powered flight, and set men free from the problem of following roads. He lived to see the Bell X-1 go through the sound barrier, and stared down the speed of light with Einstein.

It took all the years of recorded history to crack the nut of powered flight, just another 45 for Chuck Yeager to break the sound barrier, and only another 20 for Neil Armstrong to set the sub space speed record that still stands.

Orville Wright 1928-(1)

Bell x1

Neil Armstrong and X-15

And of course we will remember he walked on the moon just a little later. Some kind of rides!

Now that we know what engineers do, let's discuss how to find them.

Returning our particular problem, it remains that we have to buy the services of a few engineers to help with our plant. In particular, we will need a good electrical engineer to design our power system, a mechanical engineer will design our mechanical systems, a civil engineer may be required to design a concrete foundation, and we might need to consult with a structural engineer to confirm that what we put on the roof or hang from the ceiling is properly supported and will not collapse.

How to hire these guys can be tricky. Good engineering firms might only consist of three or four guys, or have hundreds of employees in branches across the country. Despite their collection of resumes, they will tend to have what we call a *core competence.* Many firms in Pittsburgh retain good knowledge of the steel industry; Agricultural Engineering tends to be centered in Chicago and there is good chance you can find chemical engineers in Delaware (DuPont), and electrical engineers in Schenectady (GE). The software guys are in Cupertino; the petroleum engineers are in Dallas or Houston. Mechanical engineers are everywhere. What we want for our job is a small and responsive firm. Not one where the website shows too many people standing over one drawing wearing brand new hardhats (in the office!), or standing around rotating machinery in neckties. The key reason we want a smaller firm is that it is more likely that they will talk with one another, and have good daily communications. A firm (or branch office) has gotten too large when the electrical engineers are on one floor, the mechanical engineers are on another, and the managers are on a third. They will all be communicating by e-mail rather than walking around or working around a common table. Are they having spot conversations in the hallways and over the walls of cubicles? Are they only getting together for scheduled meetings and then coming in late or unprepared? I have actually been to review meetings where a mechanical engineer has run 12" pipe right through an electrical transformer because he did not know it existed. Is the engineering manager too busy selling? That's the kind of money that just goes in the incinerator. It can be very frustrating paying for *time.* Part of what we want to pay for, as an engineering service is good communications between the disciplines.

Then, there is the question of competence or expertise. Professional engineers are those placed in positions of *public trust.* In addition to a BS degree or years of experience working under the tutelage of another PE, they will have passed an examination by a state board and are granted a stamp and license to practice in a specific field of engineering, like a barrister crossing the bar, or a doctor becoming board certified. Possession of a stamp does not; however, mean that we run right out, and hire that guy or girl. Some professional engineers have enjoyed long careers not really solving any *new* or *diverse* problems and have become *stale.* Some can be so deeply focused on a single subject, however complex, that they have become *nerds.* Some engineers have never been outside the office, and can become *obsolete.* All engineers have almost the same first two years of school when it comes to the BS degree. There will be the basic requirements in mathematics, physics, and chemistry, as well as a couple of introductory classes in each of the other branches of study. Every engineer knows that the basic element of structure is the triangle, and that gravity pulls everything to the downward. If you have a piping designer ask a question about what kind of pipe he can put an acid in, your antenna should go up. He should already know that or, find it in a handbook. If she wants to use the same pipe for nano water and tower water, and specifies 316 stainless for both, you should talk to her supervisor, and ask why. When they do their work and begin using systems of measure, be careful, we do not mix meters, and kilograms with feet and pounds.

Young engineers will have a degree of creativity, and a desire to use the latest software. Older engineers may be stuck in their ways, and think they know everything. The oldest ones will avoid obsolescence and acquire a quality we shall call *wisdom*. Engineers may come to rely a little too heavily on computer software and not enough on their own senses or grey matter. One other quality that's important are the verbal skills and how they explain things. Like a doctor with a bad bedside manner, they need not talk over your head and assume you know all of the acronyms. No one knows all the acronyms. There wasn't even LOL or WTF five years ago. The point is, you are about to get stuck with these guys for a few months. Mistakes will be made. You want them competent and confident, but not cocky or highfaluting. Look for a gang you think you will get along with. Get references, and ask those references how they are with change orders and money. At long last, engineering is a pesky product most economists would call an "experience good." The shoes might look nice, and fit nice, and be the right price, but you aren't going to know if they rub you the wrong way until you have had them on for a while.

Finally, engineers and managers do not, and many times should not, mix. Managers are charged with making money and saving time, and engineers are charged with using logic and science to make things work. The two different agendas do not go together well most of the time. If you sense tension and disagreement between the managers and the engineers during a site visit, take it as a bad sign and look into it further.

"So he (the general manager) turns to him and said 'take off your engineering hat and put on your management hat' – and that's exactly what happened,' said

Boisjoly. 'He changed his hat and changed his vote, just 30 minutes after he was the one to give the recommendation not to launch.'

Roger Boisjoly was one of the solid rocket booster engineers working for Morton Thiokol, a NASA subcontractor.

"Finally, if we are to replace standard numerical probability usage with engineering judgment, why do we find such an enormous disparity between the management estimate and the judgment of the engineers? It would appear that, for whatever purpose, be it for internal or external consumption, the management of NASA exaggerates the reliability of its product, to the point of fantasy."

- R.P. Feynman, Winner of the Nobel Prize in Physics in his Appendix to the Rogers Commission Report

Having spent some time discussing the good and bad traits of engineers, we turn back to our job at hand as we are ready to understand what it is that we expect them to do for us. Our in-house process engineer will begin to feed the basic requirements for each equipment to the mechanical and electrical engineers who will add things up and come up with a detailed design. Electrically speaking, we will need a drawing called a single line diagram which will look something like this:

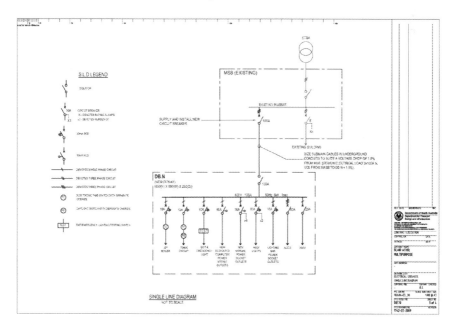

This shows the basic circuits coming from an existing busbar through underground conduits to a breaker panel that feeds 9 devices. The breakers have been sized according to the requirements developed by the process engineer. The size of the wires and conduit is not shown at this point. That will depend on the distance between the breaker and device and some other factors. In the title block of this drawing we see that this is sheet 3 of 4, so we assume that there are three more single line diagrams for this job or area. The next drawing is the conduit and cable schedule. There is no uniform protocol for this drawing, but they all should tell you everything you need to know about how many wires or cables you have, where they come from, where they need to go, and what size conduit, cable tray, or wire way they go in. A good example of a conduit and cable schedule looks like this:

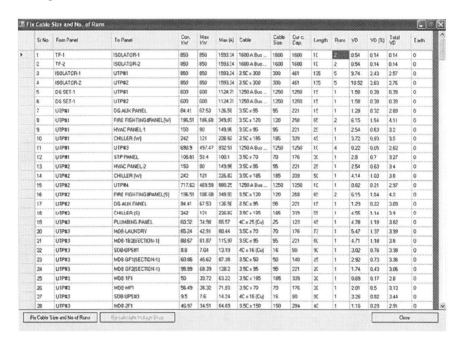

If you follow the first three columns, and work your way down, you can imagine what the single line diagram associated with this drawing would look like. TR1 and TR2 are transformers that feed 2 main Breakers, Isolators 1&2, which feed Utilities Panels 1&2. DG Set 1 is probably a back-up feed coming from a diesel generator. Lines 7-12 show what is fed from UTP-1. Lines 13-18 show what comes from UTP-2. And on and on….by now you should get the picture that when we go to our subcontractors, based on the excellent quality of this drawing, they will know the exact quantity of materials they will need to buy. We have not yet got to our *construction specifications*, which will call-out additional requirements, such as the quality and or manufacturer of the material used. A third kind of drawing which completes the electrical package, is the interconnect drawing or diagram.

For example, a good one can look like this:

Cable	Wire #	Gauge	Color	From Panel	Terminal Strip:Point	Signal	To Instrument Location	Page. Line
C654000								
	50401162							
	50401163	14	RED	CP2300	TS1:14	E-stop	ES23001:COM	5040116
	50401251	14	RED	CP2300	TS1:15	E-stop	ES23001:N.C.	5040116
	50401092	14	RED	CP2300	TS1:22	Signal	ES23001:PL	5040125
	50415181	14	WHT	CP2300	TS1:12	Neutral	ES23001:PL	5040125
	50415081	16	BRN	CP2300	AENT2300_1 Slot 16:11	+24VDC	ES23001:COM	5041518
C654001		16	BLU	CP2300	AENT2300_1 Slot 16:3	Signal	ES23001:N.C.	5041508
	50401163							
	50401164	14	RED	CP2300	TS1:15	E-stop	ES23002:COM	5040116
	50401231	14	RED	CP2300	TS1:16	E-stop	ES23002:N.C.	5040116
	50401092	14	RED	CP2300	TS1:21	Signal	ES23002:PL	5040123
	50416381	14	WHT	CP2300	TS1:11	Neutral	ES23002:PL	5040123
	50416271	16	BRN	CP2300	AENT2300_1 Slot 21:10	+24VDC	ES23002:COM	5041638
		16	BLU	CP2300	AENT2300_1 Slot 21:2	Signal	ES23002:N.C.	5041627

This is a section of an interconnect diagram for a control panel (CP2300). The whole diagram runs more than two hundred lines. This section of the drawings is telling us what to do with two conduits C654000 and C654001. Each cable has 6 conductors. Each is uniquely labeled, and colored differently as to purpose. 4 wires from each conduit (3) red and (1) white, are "terminated" or landed on a uniquely labeled terminal block inside the panel, and 2 others, brown and blue, are terminated on a slot on the process logic controller or (PLC). At the other end of the wires we find 4 devices in the form of 2 emergency stop switches. Functionally, these might be exactly the same as the float switch illustration in the first chapter when we discussed the tank automation. On the second from right, we have an integrated device/signal numbering scheme. And, on the far right we have the line number in the PLC program that receives the signal and will tell another line of code what to do in the event one of our 4 buttons is pressed or reset.

Now, it is important to reiterate that these drawings are pulled from three different jobs and were used for illustration only. It is also important to note that our Control Panel CP2300 might be part of a prefabricated equipment package coming on a skid. It might or might not already have been prewired and tested at the *vendor*. It might have been unterminated on the device side with the panel side left intact for shipping. Then again maybe you will rely on your *subcontractor* to do all of the wiring shown here. These fine points emerge as the scopes of the vendor and the contractor are revealed.

Maybe CP2300 is one of the destination panels, like line 12 on our conduit and cable schedule. Or, it could even be fed straight from a breaker in our power distribution panel. Referring back to our lessons in Chapter 1, you will continue to spend your time planning and have continuous communications with your vendor. Most of all, you need to know the scope of the electrical engineer taking care of the plant equipment. *His* cable design ends at the point where it enters the *vendor's* panel. The *vendor* is perfectly clear as to the extent of *his* shop installed wiring. The point of handoff from one scope of design or scope of work from one to another is called the *Battery Limit.* There are little demons hidden in these Battery Limits and they all come with incinerators for your money.

Here is the best published definition I could find.

Battery limit.

Definition (Credit http://www.dictionaryofconstruction.com/)

Comprises one or more geographic boundaries, imaginary or real, enclosing a plant or unit being engineered and/or erected, established for the purpose of providing a means of specifically identifying certain portions of the plant, related groups of equipment, or associated facilities.

2) A place of intersection between one scope of work and another, which is often neglected or misunderstood, miscommunicated and leads to claims for changes, lost time, and dispute.

3) A place of commercial intersection between one contract and another.

2, 3) Credit yours truly.

A good example of a battery limit in the electrical scope we have been discussing might be a note on a drawing that shows "limit of field wiring" terminating wires on the feed side of a breaker, and another pointing to the load side that says "by others"

We now turn to the mechanical side of our job. Just as seen in the electrical case, the mechanical engineer will do calculations regarding the flow rates (m*l)/t of various liquids and gasses. And, he or she will decide how big and thick the pipe or duct needs to be to transport them through the process. She will have some knowledge as to the chemical and physical properties of the fluids, and will choose appropriate materials for each application. Many types of pipe are available. Different grades of plastic, steel, and alloys of copper are in common use. We balance cost with service life and the effects of impurities due to internal corrosion or erosion. She will understand how heat is transferred, lost, and created. There will be an understanding about the selection and specifications for all types of valves, and the design and performance of different kinds of pumps. Getting started, you will have a schematic drawing for your utilities (also called a basic flow diagram). The basic flow diagram shows the direction, service, and size of the piping as well as the valves, pumps, and tanks the fluids pass through. It does not have other dimensions or provide specific locations for equipment, but it will assume we will put the cooling tower somewhere outside.

A good one might look like this:

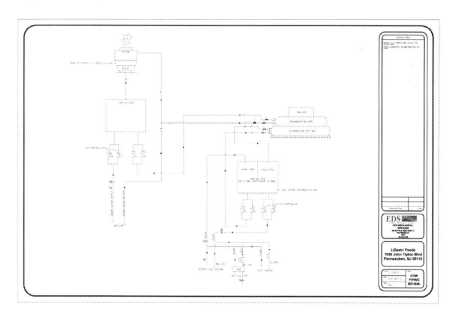

Once the basic flow is approved, the design can become a little more detailed. The capacities of the pumps, valves, and heat exchange devices emerge as additions to our equipment list. For this example, I have the list developed specifically for this diagram:

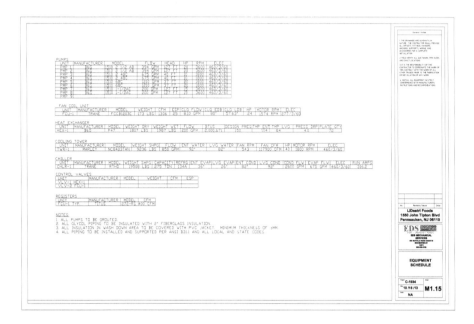

Now the equipment above is fully specified, except for the two control valves, which are still being worked. We arrive at a decision as to who should procure the equipment. We can hand it over to our procurement department, or make it the supply of the mechanical contractor who is doing the utilities scope. There are good reasons to go either way which shall be discussed in the chapter on Procurement. We collect cut sheets and begin to understand the physical dimensions and weights of our utilities gear. Our third drawing is a first pass layout showing where most things will fit in a pre-existing mechanical room. For the first time, we also show the utilities piping running to the process equipment. The mechanical room is shown in the upper right, and the process equipment is shown, all to scale.

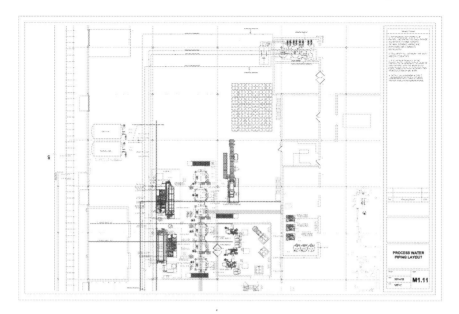

This segment or scope of work is almost ready for bidding the installation work. The new equipment has been specified, and the pipe has been sized. We can estimate the length of the pipe in the ceiling and decide the support intervals. The vertical "drops" from the horizontal headers can be located, and the battery limits can be made precise. For example we will prepare instructions stating "the scope of work on this drawing shall terminate at a two inch flange located on the process equipment PR-MT See drawing… XXX for exact location." Flange shall be made up and bolts supplied by contractor. Pretty hard to misunderstand that one.

By this time, it has taken three months to get the ball rolling. Most of the equipment has been specified, some even selected. Files have been established. The basic engineering is under way with our process mechanical and electrical engineers, who have a good idea of what they need to do. They are in positive daily communications. Instead of questions, they come to meetings and call on the phone with better ideas about what was talked about last week. The rest can be done in-house. We begin to take charge of everyday. Our equipment list is nearly complete. There have been enough discussions with prospective vendors that we have an idea of who we will ask to make a commercial offer. We are not being *sold* anything yet. We are developing our judgement and using our *senses*.

We have made that one big mistake with the one-off machine that was let to bid with no success. Probably $50,000 was wasted because we did not know what we were doing. It is still considered by upper management to be 2 or 3 million beyond a reasonable budget. We found the old file on that machine, and we believe management. We find the old actors on that job are gone, and we will have to do it with our own wits. We do not point fingers at the earlier waste. We have worked the solution, and have parsed the machine to a list of hardware, purchased, and machined parts. We have developed a *path forward* with Bob.

So, we can now turn to chapter 3. We need to make a clean pivot to Procurement. This is yet another subproject, or segment of our overall job. It is wrought with new actors and deadfalls. We need to move quickly, and yet with a great deal of care.

CHAPTER 3

Ideas upon the procurement of unique, expensive, and ordinary industrial goods, with a side lesson on Power and Politics in organizations.

As we turn the chapter to the procurement of our off-the-shelf items, specialty equipment based on performance specifications, and our complicated, one off machinery it is necessary to understand that we have been thrust upon some actors upstairs in the procurement department, and they already don't like us.

When we came on board, our new bosses in upper management told us that the easy - and we inferred - politically correct way to get our one-off complicated machine built was to turn it over to the powerful corporate procurement department. Not yet having a feel for our own position and the abilities of others, we did what we were told. Four weeks later the non-responsive bid results were right back in our lap as discussed in Chapter 1 and this subproject has gone straight to the hunt for the guilty. You, the outsider, have been temporarily assigned as the "project manager." It does not matter if you have invested 20 years of your life in the same corporation. If you hail from another plant or are an independent contractor, you will be gone when the job is done. The procurement department does not need you for anything. In fact you have inadvertently invaded their fiefdom. Upper management has not set the proper tone to support the project. While they have delegated budgetary and schedule responsibility to you, they have set the stage for failure by placing you in a position of lower *rank* by giving you a temporary office in the boiler room, noisy, and far from a toilet or water. There should be a project organization chart with the PM or CM at the top, and a dashed line to operations, maintenance, procurement, accounting, and finance to emphasize these line functions are in position to *support the project.*

And so, while you are charged with spending $50 million in complicated and highly visible non-routine project dollars over the next two years, the procurement department in a hypothetical bottling plant will spend $5 million on predictable and routine things like blow molded plastic bottles, water, sugar, and cardboard. They exist to beat suppliers on price, then write the contracts for five years. In the meantime, they approve invoices that are the same every month, checked off a barcoded receiving system, and the only thing they fear from the inside is an unscheduled audit. They get to knock off after lunch and then go play golf with the boss and one of the suppliers.

They sure don't want anyone to upset that applecart.

Further,

The procurement department remains in the fortified and politically entrenched position next to the corner office. They chat with upper management over morning Starbucks. They will blame you for the bidding failure, on the first go at the custom machine, behind your back and start to get away with it. They are *accustomed* to the way things are, and invested heavily in the *status quo*.

Hunter S. Thompson liked the Latin phrase *res ipsa loquitur* - "it speaks for itself". There is no direct evidence that procurement and management are misbehaving or possessed of ill intent toward you or the project. Everyone, without a doubt wants the project to succeed. They will need to buy even more plastic bottles and cardboard. They just did not know what to do with this expensive and fragile basket of expensive eggs handed down by the board of directors. It just remains that there is a new sheriff in town, and the first thing he did was start to upset the applecart.

ASIDE

Once, on a job in Decatur Alabama, a good guy and excellent millwright named Perry Powers told me "I don't care one shit where you came from or how good you do by me. If you're from the other side of Tennessee River, you're a *Yankee*."

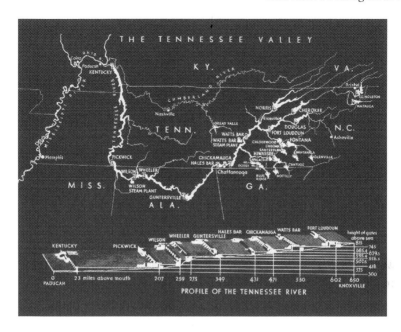

I said to Perry:

"You're pointing at Huntsville. When I did a job in North Carolina the pipefitters were calling me a *Goddamned Yankee*"

Perry said:

"No, no, Dave, they got it wrong. The only Goddamned Yankees round these parts are the ones that come here to stay."

END OF ASIDE

The germ of the conflict with the purchasing department was laid long before we arrived on the scene. In fact, it is one of the elements of human behavior. We like things the way they are, and we dislike change. Victims of learned behavior, we are even taught that there is one and only one correct way to load the dishwasher. The project may be the biggest thing to happen in this plant in 20 years and people are nervous. Fortunately, our project manager, that goddamned outsider agent of change, comes on the scene with a tool to slice out the germ before it festers. If he does well, everyone will come to like him. The good project manager *teaches* procurement and management what to do with the basket of eggs. She *leads*. The mediocre project manager will be released the moment there is light at the end of the tunnel. The excellent project manager just makes it look easy; he wins people over with his smile and confidence. They will be sad to see him go.

And we stop complaining about the purchasing function in general by noting that they can be the corporate savior. Think Southwest Airlines, with the futures contracts on fuel in the early 2000's, and reaping enormous economies of scale by flying 737's only. Or was the play on fuel futures just another form of gambling? The contracts expired about the same time oil prices started to crash in 2014.

The first thing we do is assert our rank. When men meet men or women meet women, I have come to the conclusion that the response can be rather canine. Dogs are instantaneously excited, take a quick sniff, and then decide to have a fight or go run and play. Rank is very important in our temporary organization because we neither have time to fight or run and play. And the title "project manager" itself can be problematic. There are some out there that have the title of project manager and handle 10 or 20 simultaneous projects ranging in price from $5,000 to $20,000, but are not allowed to write a purchase order of more than $100 without higher consent. Others will have the same title, and be responsible for a $10,000,000 job but it all comes in the form of materials and might be as simple as milling and repaving a parking lot. What I am driving at is that the title of project manager can carry a very broad range of responsibility and accountability, and organizational rank is derived therefrom. At the top we would have guys like Teddy Roosevelt or John Roebling who built the Panama Canal and Brooklyn Bridge respectively. Or even Eisenhower, who spent most of his time *planning* D-Day. We want to be like these guys. We will not participate in finger pointing. We move stepwise and purposefully in a positive direction. People in the plant *need project leadership* and will come to appreciate it, even if it comes from an outsider. Think Lafayette.

Our rank is asserted by sending our project engineer upstairs with the list of all of the purchased parts that we have parsed from the drawings. Granted, we do not need most of them right away, but we have the requirements all on one list; part numbers, manufacturers, recommended vendors, web links to the vendors, phone numbers, and names. We have handed over a complete and specific requisition. This part of the job has been de-mystified to the point where they should be able to handle it and they will come to appreciate that you have taken care of all of the *non-routine* aspects of the business. About the only thing that will need to be done is to make sure that our tagging requirements make it on the PO. The point of sending our project engineer upstairs with everything is so that he can have a face to face meeting with the buyers in your absence and it will be clear that on this level of the project you have delegated the business. You do not, personally, ask one of the buyers to help you, because that simple and otherwise harmless request would place you in a rank among the buyers, and a notch below the purchasing manager. A few hours later, having seen what you delivered to the buyers, the richness and depth of all the information, and the fact that you sent the project engineer, the purchasing manager realizes that you are not there to upset the applecart after all. In fact he gives you a call to make nice with one or two trivial questions, and heads out to golf. He is relieved because you have led him back to stage 1. You are no longer his prey in stage 4.

In the meantime, we construct the bid form and instructions to bidders for all of those remaining machined parts. Recall our list looks like this:

1 Line No.	2 Description	3 Quantity	4 Drawing Number	5 Machine Number	6 Tag	7 Unit Price	8 Extended Price
1	Base Plate	3	P-DS-BP 1-2-5	P-DS-1-2-5	P-DS-BP-1-2-5	$ 2,000.00	$ 6,000.00
2	Base Plate	2	P-DS-BP-3, (4OPP)	P-DS-3-4	P-DS-BP-3-4	$ 1,500.00	$ 3,000.00
3	Transmission Gear	5	P-DS-TG	P-DS-1-2-3-4-5	P-DS-TG-1-2-3-4-5	$ 500.00	$ 2,500.00
4	Transmission Gear (2)	1	P-DS-TG2	P-DS-3	P-DS-3-TG2	$ 200.00	$ 200.00
5	Transmission Gear (3)	1	P-DS-TG3	P-DS-4	P-DS-4-TG3	$ 200.00	$ 200.00
6	Main Shaft	3	P-DS-MS-1-2-5	P-DS-1-2-5	P-DS-1-2-5	$ 500.00	$ 1,500.00
7	Driven Shaft	6	P-DS-DRS-1-2-5	P-DS-1-2-5	P-DS-DRS-1-2-5	$ 400.00	$ 2,400.00
8	Main Shaft (2)	2	P-DS-MS2-3-4	P-DS-3-4	P-DS-MS-3-4	$ 500.00	$ 1,000.00
9	Driven Shaft (2)	4	P-DS-DRS	P-DS-3-4	P-DS-DRS-3-4	$ 400.00	$ 1,600.00
10	Frame	3	P-DS-FR-1-2-5	P-DS-1-2-5	P-DS-FR-1-2-5	$ 600.00	$ 1,800.00
11	Frame	2	P-DS-FR-3, (4OPP)	P-DS-3-4	P-DS-FR-3-4	$ 650.00	$ 1,300.00
11-499	Blah	Blah	Blah	Blah	Blah		
500	Name Plate	5	P-DS-NP	P-DS-1-2-3-4-5	N/A	$ 150.00	$ 750.00

And I'll reiterate that it does run to 500 lines. A substantial amount of work to bid, track, and receive.

For this procurement, the Instructions to Bidders (IFB) and The RFQ (Request for Quotation) can be one in the same document or separate, or, if we are undertaking a more simple procurement, the IFB may not be necessary. In our case, the initial procurement failed because the vendors did not know what to do and/or did not have enough time, motivation, or desire to figure it out.

We will proceed by preparing the document with an RFQ as a preface to the IFB In a single document:

June 13, 2015
Request for Quotation:

Jim's Tool & Die, Bob, and three others.

Dear Jim,

Please find attached a substantial set of drawings for machined parts required for our project in Parsippany New Jersey. The drawings represent machined items required for the construction of five drive stations. This procurement is for the machined parts only. Hardware, purchased parts, assembly services shall be by others. Plating, painting, and heat treatment, where indicated shall be included within your scope.

The schedule for the procurement is as follows:

Bids Due July 20, 2015
Purchase Order / Notice to Proceed July 27, 2015
Deliver Base Plates September 1, 2015
All Other Items Not later Than October 1, 2015
Commercial Terms as Agreed.

Instructions to Bidders:

We have prepared a bid form for your convenience as an excel spreadsheet. Each line on the spreadsheet corresponds to a drawing in the package, and the quantities have been indicated. Opposite hand requirements are also called out. Formulas extending your unit costs to extended costs and total costs have been inserted. It shall remain your duty to check the math.

We fully understand that the scope and schedule of this project will require the services of multiple shops and do not expect full coverage from any one respondent. You are instructed to bid only those items within the immediate capability of your shop. Subcontracting shall be limited to painting, plating and/or heat treatment. Do not bid more than you can deliver by October 1, 2015. Late deliveries are unacceptable and may be rejected. However, a delay on our part in issuing the purchase order shall extend the due date on a day by day basis if necessary. Assembly and sub-assembly drawings are issued for reference only. All assembly shall be by others.

You may be awarded all, some, or none of the items quoted. We anticipate working with multiple shops, and intend to order from the shop offering the lowest price for each item. We do not anticipate any further negotiations on price after the bids are received, except in the event of a tie. Please return the bids as "best and final". Once the initial awards are made, we anticipate a second or even third round of bidding in rapid succession until the requirement is closed.

All uncoated machined surfaces shall be protected with a suitable corrosion inhibitor. Each item shall be tagged with the number shown on the bid form.

Use an industry standard #8 tag, McMaster # 1547T26 or equal. Write with sharpie or otherwise water insoluble ink. Parts without a tag shall not be received and will be returned at bidder's expense. Please do not ship any single pallet or skid which cannot be picked with an ordinary 4000# forklift.

Requests for information shall be by e-mail. Reference the drawing number. Only one question per e-mail please. Answers shall be made as quickly as possible and returned to all bidders by e-mail in adobe pdf format. Verbal questions shall not be answered unless the question is unique to your operations.

Freight shall be pre-paid and added to your invoice. You may ship for convenience within the aforementioned constraints. Should there be any conflict between the requirements in the fine print of the purchase order and these instructions, these instructions shall prevail. If there are any discrepancies between the quantities shown on the drawings and the quantities on the bid form, the drawings shall prevail and an appropriate change to the extended price will be made.

We appreciate your time in quoting this project, and look forward to the receipt of your offer.

Sincerely,

***

Now, the de-mystification of this piece of work is nearly complete. The Instructions have made it perfectly clear what we expect the shops to do, and what we intend to do with the results. We fully expect that any shop not fully booked for the year will provide some kind of response. And now that the guys in the purchasing department are beginning to like us, we wait a few days until most or all of the purchased parts are put on order and send our project engineer back upstairs with RFQ/IFB we just created and, *for information only,* let them know what we are doing. When the quotes come in, we do our evaluation, pick out what from whom, and then do the rest by way of a standard purchase requisition. The buyer's job has been reduced, to the extent possible to a few keystrokes, about as simple as buying a pallet of cardboard, or releasing a shipment against a blanket order. Remember lesson learned #3 in chapter 1. We have just recognized and taken one of the hard things, and did it by ourselves. We have avoided making enemies when people were predisposed not to like us.

The next mistake we avoid is another unwanted surprise. So on the fourth day after the package goes out, we proactively call, *by telephone,* each of our shops to confirm their receipt, and gauge their initial interest. If shop number 1 wants to, but cannot respond, he probably knows of another well qualified shop down the road and by following up we can identify them quickly.

We make *human* contact instead of just sending a package out by-email. By now, everyone knows that even this type of document might get caught by a spam filter, and there is always one or two guys out there who has an internet service provider that cannot handle a big file. There are also the cloud services where everyone gets a password and can check for RFI's and other project documents on-line. These services are evolving, and can be quite useful, but at the time of this writing in June of 2015, I am going to say that the jury is still out because there are not enough protocols in place for security and document control. Presently, I prefer to stick with telephone, e-mail on a one on one basis, and a pdf for documents requiring barriers to unwanted modifications. Let's put the one-off procurement to bed by imagining that we will assemble the drive stations using labor we get from Bob's Shop. He has millwrights with enough skill, and the labor content of the assembly is not that great. We imagine 3 guys can get it done in three weeks per machine (45 man-weeks) total or 1800 man-hours. It is decided; however, that the work will be done in the plant instead of Bob's Shop. There is a small plant crane and a forklift at our disposal, and they can be either dedicated for our use or we call for a rental. We do this because having spent so much time with the drawings, we can imagine the assembly process in our head. We will be right there with them on-site to help and answer questions. This part of the job will get done on a time and material basis, because we know the risk better than anyone, and there is no need to pay for that risk in the form of a lump sum arrangement on the assembly. We also stand to save a few thousand dollars that would have been spent on shipping the finished machines.

Hardware is yet uncovered, but we know we will need enough that it will make sense for the store to come to us. We do not need the millwrights running to the store because we keep a short supply. Hardware commonly comes in package quantities of at least 5 pieces, and as the bolts or other fasteners get smaller, the package quantities go up. 10, 25, 50, and 100 are common package sizes. We take a little time and do a take-off from all of the drawings and order everything from an (ideally) local industrial supplier, such as Fastenal. Close is good enough; the Millwright foreman is instructed to let you know when things are getting low – it is going to be his problem if work is halted because we run out of a nut or bolt on his watch. We also ask the supplier to come in once a week to do an inventory and requisition for resupply. We set up bins, label them clearly, and build some shelves from cinderblocks and planks. More on site management and controls will be discussed later.

Now, let's take a look back for a moment, and look at some scheduling fundamentals. First we made an effort to understand what are called the "long poles" - equipment that might take a year to get built is treated differently than equipment taking 4 or 6 months. Equipment taking a few weeks is different yet again. Each time we talk to a vendor, we have a casual discussion about the lead times, so it begins to stick in our head. Then, there was the mission of discovery, where we spend effort up front coming to know the unknowns. This is the basic engineering portion of the project where we discovered the need (or not) for a new compressor or boiler. The framework of our schedule begins to emerge. It goes without saying that the longer it takes to get our hands on something the sooner we need to get it together AND on order, because:

AXIOM 9

THE LONGER THE LEAD TIME ON CAPITAL EQUIPMENT, THE GREATER THE POSSIBILITY OF UNFORSEEN DELAY, CHANGES, OR PROBLEMS, ALL ELSE BEING EQUAL.

COROLLARY TO AXIOM 9

A SUITABLE AMOUNT OF SLACK SHOULD BE BUILT INTO THE SCHEDULE TO CONTROL WHAT GOES ON WITH AXIOM 9

And of course, our early focus was on the one-off machinery, because all we knew was that it was a month behind schedule before we got our hands on the problem.

Returning to the balance of the equipment, we have several criteria to consider. Initially, vendors will return their quotes, there will be revisions, and then refinements. One of the keys to the successful procurement is to keep the technical specifications and engineering development separated in thought until you are ready to cut the PO. Take some time to look at the future costs of operation and maintenance among your alternate choices. Does one machine offer automation to the extent that there is an appreciable difference in the labor required to run the machine? How many hours is required for maintenance each year? Finally, to the extent that things can be quantified, we run a net present value on our alternates:

$$\text{NPV} = \sum_{t=0}^{n} \frac{(\text{Benefits} - \text{Costs})_t}{(1 + r)^t}$$

where:
r = discount rate
t = year
n = analytic horizon (in years)

Since we are now making a comparison among alternates, we can assume the benefits (outputs) are equally specified. We also assume that any future income from salvage is far down the road and trivial. The analytical horizon is the service life of the machine, and the costs are our best estimate of operations and maintenance expenses over time. The cost of the investment is taken in year 1, and each machine is charged the same discount rate.

As we are dealing with costs only, each case returns a negative number; the proper choice is the least negative of the alternates. Now if you wish, you can take this cost of ownership and add another alternate, which would be the cost of a lease if one is available.

The chapter on scheduling will go into detail about what should come, first and last, when and how. For now, we make the observation that our equipment quotes all have a delivery clause that says something like "delivery 6-10 weeks after receipt of order." The thing to do at this point in our project is look at the equipment piece by piece, make a determination of when *we* want it, and write it into the purchase order. *Equipment shall be ready to ship on April 20, 2015.* Include in the PO so many days of storage on vendor's site, without additional charge, in case there are some unforeseen delays on the jobsite beyond your control. Make sure you require a description of case dimensions and weights as soon as it is available. Case dimensions and weights become another one of those consolidated lists that will become one of the bid documents for the rigging and setting portion of the job. We take care not to confuse pounds and kilograms (mass!), feet and meters (length!), or days and weeks (time!) It goes without saying, that we want the rigging contractor on-site and ready with the right gear the day before the equipment starts to arrive.

We revisit and reconfirm our Battery Limits. Here is a good example of why we continue to pay close attention. One machine cut sheet says the machine requires 20# per hour of steam at 32 psig. Our boiler kicks out more than enough steam at 50 psig. The machine requirement is sent over to the mechanical engineer and he knows we will need a standard regulating set-up to reduce and control the pressure. He comes up with a design that results in a drawing that looks like this:

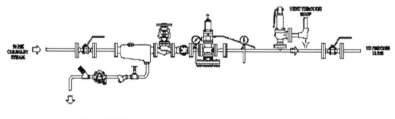

1 CULINARY STEAM PRESSURE REDUCTION DETAIL
NOT TO SCALE

The supply of the labor and material for this reducing station becomes part of the scope of work for the mechanical contractor. Let's suppose we have four of them, and the cost for each one is $10,000.

It only becomes evident much later, when we receive the equipment along with the vendor's engineer and his 500 page equipment manual, that there is already a functionally equivalent reducing station (PRV) built in to the machine. Yes, the vendor's utility requirement, *(at a minimum)* is 20# at 32psig, but the machine could have handled much more pressure at the steam inlet. Now, let's look at the *timing* of our discovery. If we discover the redundancy when we are in the process of bidding the job, we write an addendum eliminating the work, and the only damage done was the amount of time the engineer spent designing this unnecessary redundancy. If the discovery is made after the job is let, but before the material is delivered, we will be able to go to the subcontractor with evidence of the material cost due as a credit. Another issue is that leverage against the credit requested for the labor will be on the contractor's side and he or she will offer 20 cents on the dollar. If we make the discovery after materials are received, but before they are installed, we pay a restocking fee on what is returnable. Finally, if the discovery is made after the construction is complete, we might as well scrap it in place.

These scenarios, in what we find out and when can play out very quickly. Maybe a week, maybe a little more. The little demons in the battery limit just got the whole $40,000. If we had done a better job in planning we might have gotten our hands on a flow schematic ahead of time, and avoided this situation altogether. There are other administrative controls available for this problem when we reach the chapter on bidding for construction. We might ask our project engineer to take some time and review each and every battery limit on the project. They can run to the hundreds and be simple – like a two inch flange, or more complicated than the case above.

***

Returning to our drive station project, it is now July 20, and having fully understood our requirement and in accordance with (IAW) our IFB, we receive 5 responsive bids. At least two or three for about 350 parts, only one bid for 100, and none at all for the other 50. We need to decide how to deal with the results. New questions arise, and some answers emerge like a high school science project.

If the pricing on the 350 items is tight, we might go straight to purchase. Two close bids are enough to release the work. Does the same shop come in third place most of the time? Maybe he is at a cost disadvantage. Where we have only one price, are they scattered amongst the five shops in a similar pattern as the 350 and do they seem fair, or was there something unique about the 100 parts – maybe very large borings, or smaller custom gears? It might make perfect sense that only one shop had the tooling to make them. We take our results for the 450 parts and make a Notice of Potential Award (NPA). We may or may not decide to release all of the results to everyone, or just what each shop is awarded. I usually prefer to release everything because it keeps things transparent and the shops may appreciate what they see, win or lose. The NPA is just another spreadsheet, which we parse out in 48 hours, and send over by e-mail. Then, we pick up the phone and schedule a site visit.

We do not care about what the shop looks like from the outside. It may be 100 years old next to an abandoned steel mill, it may be on a downtown backstreet, or out in the boondocks. We step inside, and might discover that they are still running MS DOS and printing with dot matrix printers. They still have a fax machine; although it is gathering dust. There can be paper all over. You sent them the drawings in electronic form and they spent hundreds printing at Fed-Ex Kinkos. They are marked up with notes and expected hours of machine time. The office is clean and the guys coming in from the shop floor wipe their feet. Most of what goes on in this office is of little concern to you, not like that somewhat dysfunctional engineering office we discussed earlier.

What we do want is a tour of the shop and we want to see which machine will be used for each part. How many of each machine do they have available? Did the shop foreman participate in the bidding – did he agree with the schedule? If the shop foreman and the guy who did the bid did not work together, you might have a problem. Did they quote the plating and painting as a prime contractor? Look at the plating tank and make sure the biggest part can fit in. Look at the paint shop and make sure you see more than a bunch of half empty cans and dried up brushes. See if they can show you a sample painted object. If you are not convinced they will do a good job, ask them to submit a blank. When upper management comes around to look at things later in the job, the only comment they may make will have something to do with the *appearance* of the paint. There would be no clue as to whether it is just some oil-based enamel, or a marine grade epoxy.

If all goes well, we leave the shop with a good feeling – they want the work, they can do it on time. Look them in the eye and shake hands. On to the next. The NPA goes over to the purchasing department as a requisition, and in a few days the orders are placed. The 50 remaining items are sent out for re-bid, and our 5 shops, knowing us a little better, and that they all got *at least some* of the job, return new bids for all but five.

The last five are special. Why? We ask all the shops. Maybe there was something special or exotic about the material. Maybe there needed to be a custom mill run of a specialty steel, or a diamond tool to machine a ceramic.

Nothing that was done before should be impossible to do today. There is some solution and when looking at the five drawings side by side and in isolation. The answer is before us....we find that the material is obsolete. Still available, but hard to locate. Based on a series of communications, and questions, we finally come across an *expert* and find a workable alternate as to material or design.

Thus, we close our chapter on procurement and you know what to do. The initial problem of the procurement of our one-off machine has been solved. The new price appears to be about $2 million. Upper management gets to sit back and say they just knew they were right. Now we have a total and accurate budget for all of the equipment and have compared it with the estimate in the original budget. If it is lower, we are in good shape, if it is higher we point it out to upper management and open discussions about a potential increase in the equipment budget line item.

Maybe we run another project NPV at this point. We imagine it remains positive and after a rather extended period of time, we have satisfied Axiom 1.

The lessons we have learned in this chapter are as follows:

- The major capital equipment will be scheduled to come when we want it, and not whenever the vendor tells us it will be ready.
- We won over the purchasing department by making their participation in the project fit their daily routine to the extent possible.
- We further developed the net present value analysis as a tool to be used in many stages of the project and used it to justify the purchase of a higher priced machine that will cost less in the long run.
- We came to see that not all offices should look the same. We want our shops to spend their money on state of the art machinery and don't care if they are still using red pencils and a DOS version of Wordperfect.
- We completed the theme of de-mystification by getting down to our last five drawings, where we discovered the natural element of the problem and found a way to solve it.
- We demonstrated the importance of understanding the Battery Limit, and how misunderstanding this junction of responsibility can lead to a costly error.

CHAPTER 4

How to Pre-qualify and Shortlist Subcontractors.

There is, of course, a very wide variety of contractors offering just as many skilled services as are needed to get work done. There are also widely recognized trades or crafts. We will refer to a craft worker as a skilled manual worker, proficient or having mastered his craft. We will also need to gain an early understanding of the differences between a craft union, and a labor or service union. Everyone works. The easiest way to differentiate between a craft and service union is that craft workers work on jobs that are temporary in nature, ie: construction or maintenance turnarounds and will generally work for many different employers over their career. They pay union dues to the craft union and, in return enjoy collective bargaining power negotiated with a contractors association and derive pensions and healthcare benefits from the union. They will work from a local hall within a territory, or travel if desired and work from any other hall having work available. On the other hand, service or labor union workers get to stay in the same place for a long time – police, firefighters, hospital workers, teachers, etc.

Craft unions usually require a four year apprenticeship, where there is a structure for the entry level worker to learn his or her skills through a training center and on the job under the tutelage of a Journeyman, one who has mastered his craft and is interested and capable of teaching.

The Union Contractor is typically one who has made an investment in tools and equipment to support the work of one or more craft, and will pull his labor from the union hall as needed. They will be able to handle a wide variety of scopes of work within the craft they support.

Then there are non-union contractors that may be a small or even very large family business or corporations. Some are satisfied working within a limited range, servicing more local requirements: heating and air conditioning, plumbers, some painters and carpenters etc. There are electrical contractors that focus on residential and service work. Asphalt and concrete never get poured too far from the plant. Union Labor generally gets paid more than non-union on an hourly basis, but in theory they are more subject to lay-off when things get slow.

There are also the General Contractors. Some might have a relatively small permanent work force, an owner, estimator, superintendent, and administrator. Here we think of home builders, restaurants, big-box retail, and shopping mall fit-outs. These guys will take a set of drawings from say Hampton Inns, and parse the job down to its basic scopes. They might wind up with many subcontracts:

1. Earthwork and Grading, Site Drainage.
2. Asphalt
3. Line Painting
4. Framing
5. Moisture Barriers
6. Exterior Stucco
7. Drywall, Rough-in and Finished
8. Electrical
9. Plumbing and Fixtures
10. Fire Protection
11. Acoustic Ceiling
12. Finish Painting
13. Lighting
14. Finished Carpentry –Millwork & Cabinets
15. Glazing – Windows
16. Air conditioning
17. Roofing
18. Carpeting
19. Tile & Terrazzo
20. Swimming Pools
21. Elevators

The foregoing is just to give you an idea of how many different actors might come into play depending on the strategy of the GC. He may have experience with a multitude of subs within a certain area and take only one or two bids, but generally what he does is sub out everything and the have a direct hire superintendent on the job whose goal it is to get everyone working in peace and harmony. The project might seem complex, but the work gets so repetitive and routine that things are easily de-mystified. Within the specifications for construction, the work has been divided into standard divisions and the GC can just tell the electrical contractors they are responsible for everything within division 16, the plumbers do division 15, and the stucco workers do 04. Through 2004, the Construction Specifications Institute defined exactly 16 divisions of work.

- Division 01 — General Requirements
- Division 02 — Site Construction
- Division 03 — Concrete
- Division 04 — Masonry
- Division 05 — Metals
- Division 06 — Wood and Plastics
- Division 07 — Thermal and Moisture Protection
- Division 08 — Doors and Windows
- Division 09 — Finishes
- Division 10 — Specialties
- Division 11 — Equipment
- Division 12 — Furnishings
- Division 13 — Special Construction
- Division 14 — Conveying Systems
- Division 15 — Mechanical
- Division 16 — Electrical

After, 2004, the list was expanded to 50 divisions. I'll not trouble the reader by going into that list. 16 was probably the right number of divisions to describe most all of the work in a useful manner. We just close by noting that beyond the small GC's there are very large ones with global reach, such as Bechtel, CH2MHILL, Shimizu, Kajima.

In setting the stage for this chapter, I have attempted to present a brief picture of the variety and types of construction to demonstrate a little about how labor works – union or non-union, and that there are family owned businesses, partnerships, mom and pops, and multinational corporations competing for and doing very similar work. We are talking about annual revenues from 1 million a year to more than a billion, and permanent employees from just 5 to more than 50,000. On our hypothetical project we have been developing, we do not need 10 or 15 subcontractors to get our job done, but we are going to need 5 or 6. Assume we are working in a new geographical area and we don't know the players yet. The overarching question to be answered in this chapter is how to decide which contractors we want and, just as importantly *do not* want to work with. The question is fundamental to our project. The correct decision will lead to a good job down a smooth road, finished on time and on budget. The incorrect decision will lead to a type one or type two error.

Before presenting a model of how the decision is made for our particular project, let's spend a moment and learn a little more about craft labor, so we understand a little more about what they do and how they are organized. This will be useful to the reader for other subsequent projects beyond the one demonstrated here.

The following is a fairly complete glossary of trades in construction.

- Boilermaker - works in nuclear and fossil power plants, shipyards, refineries and chemical plants, on boilers, pressure vessels, and similar equipment.
- Carpenter - a craftsman who performs carpentry, building mainly with wood. Among carpentry's subsidiary trades are those of cabinet maker and millworker, cladder/sheeter, framer, joiner, and roofer. Carpenters unions usually include drywall installer/lather, flooring installer, pile driver, millwright, diver, and diver tender.
- Carpet layer - one who specializes in laying carpet
- Electrician - specializing in electrical wiring of buildings and related equipment. Electricians may be employed in the construction of new buildings or maintenance of existing electrical infrastructure. Power Line Technicians, High voltage line and substation construction and maintenance trade. Electricians also do the telecom cabling, Ethernet and fiber optic data cabling.
- Elevator Mechanic - installs vertical lift and transporting equipment
- Fencer - a tradesman who builds fences
- Glazier – cuts and installs glass usually in aluminum extrusions in high rise buildings and storefronts. May be included with Painters
- Heavy Equipment Operator aka Operating Engineer - a driver and operator of heavy equipment used in engineering and construction projects. There may be many special function titles, such as Bargeman, Brakeman, Compressor operator, Elevator operator, Engineer Oiler, Forklift operator, Generator, pump or compressor plant operator, Signalman, Switchman, Conveyor operator, Fireman, Skiploader

operator, Helicopter radioman, Boring machine operator, Boxman or mixerman, Asphalt plant engineer, Batch plant operator, Bit sharpener, Micro tunnel system operator, Pavement breaker operator, Drill Doctor, Drilling machine operator, Rotary drill operator, Canal liner operator, Canal trimmer operator, etc.

- Insulation Installer - Includes application of all insulating materials, protective coverings, coatings and finishes to all types of mechanical systems. Also Hazardous Material Handler. Asbestos workers also fall in this category.
- Ironworker - (or *steel erector*), erects or dismantles structural steel frames. Structural steel installation is usually crane-assisted. Workers rely on mobile, elevated platforms or scissor lifts. Ironworkers bolt the steelwork together using various tools, power tools and manual tools. Metallic Lathers may be included in this category. Not to be confused with Steelworkers who make steel in the mill from raw materials
- Laborer - a skilled worker proficient with pneumatic tools, hand tools, blasting, smaller heavy equipment. Laborers may also assist other tradesmen.
- Landscaper - a tradesmen who specializes in landscaping.
- Mason - a tradesman skilled variously in brick and blocklaying, concrete finishing (the placement, finishing, protecting and repairing of concrete in construction projects). Also stonemason, marble setter and polisher, tile setter and polisher, terrazzo worker and finisher. A Hod-carrier is a subsidiary trade, (nearly extinct in the US), one who brings bricks to the bricklayers and blocks to the block layers.
- Millwright / Rigger – Unloads moves assembles and installs various industrial equipment. Usually works through the Carpenters hall.
- Pile Driver - a tradesman who installs piles, drills shafts, and constructs certain foundation support elements
- Plasterer - a tradesman who works with plaster, such as forming a layer of plaster on an interior wall or plaster decorative moldings on ceilings or walls.
- Plumber - a tradesman who specializes in installing and maintaining systems used for plumbing, heating, drainage, potable (drinking) water or small-sized industrial process plant piping.
- Pipefitter - (or steamfitter), a person who lays out, assembles, fabricates, maintains, and repairs large-sized piping systems capable of enabling high-pressure flow.

- Sheet Metal Worker - installs HVAC ductwork, industrial ducting for dust collection, vapor collection exhaust stacks and related work
- Fire Sprinkler Installer, installs fire sprinkler systems. Usually held as a distinct trade from the Plumbers and Pipefitters.
- Safety manager / Safety Officer – Usually a safety scientist or engineer who oversees and coordinates safety affairs across the craft on the site.
- Site manager – A senior Tradesman or Engineer who coordinates all inter-craft activities on the site. More often this is a frontline, non-union management position.
- Rodbuster – an Ironworker who positions and secures reinforcing bars and mesh used to reinforce concrete on construction projects.
- Teamster - generally drives a truck or vehicle of some kind, but may also work on material handling and storage, works to support any other craft. The union was founded in 1903 when men would drive a "team" of horses. Hence the logo:

- Waterproofer – Applies waterproofing compounds to basements, foundations, and concrete.
- Welder - a tradesman who specializes in welding.

As stated above, we are undertaking this project in a geographical area where we have no experience. Our search for the subcontractor starts from scratch, there are no walk-ins or trusted referrals. Beginning with a blank page, the effort from the first day of our search until the day we are ready to execute the contract will take about 3 months. Three months may sound like a long time, and we could possibly arrive at good choices with less time, but presently, the reader is referred to axiom 2 (65% Planning).

ASIDE

Consider the purchase choices we make in our daily existence. Unless we are among the very wealthy, most of what we buy is based on price, even to the point of irrationality. We might drive a mile out of the way (at the IRS rate of $0.57 per mile) to "save" $.02 per gallon when we need to fill our 20 gallon tank. We buy one of something we don't really need to get another one for "free". Even when making a major purchase of an experience good such as an automobile, we are bombarded by so much lifestyle advertising that people wind up driving the Hummer with the off-road package and heated mirrors in places like Miami. Now, even though we have company funds at our disposal, we must attempt to cast aside both irrational learned behavior and *keep the contractors selling effort at arm's length.*

END OF ASIDE

There is a process to uncovering the suitable subcontractor by answering a few basic questions. We note that these questions must be answered to our satisfaction *before* we invite anyone to submit a commercial proposal. One of the worst errors we can make at this stage is rushing to a contract with an unrealistically low priced, unqualified subcontractor, or one who is buddies with someone upstairs.

The new unknowns, which we must come to know, are briefly:

1. What is out there?
2. Are they interested in our project?
3. Is the project a good fit for the subcontractor?
4. Does the sub have verifiable references?

The basic idea is to cast a wide net and filter the catch.

We begin by searching ThomasNet or google: "Electrical Contractors" "Mechanical Contractors" "Construction Contractors" "Rigging Contractors", and begin to get a feel for what is out there. If we are in a metropolitan area, we might easily locate 10 or 15 prospective contractors for each division of work within a 50 mile radius. In the mid-West, we might need to look at a 200 mile radius to find a fewer number of prospects. Let's proceed from the point of view that within a couple of days we have located 15 prospective subcontractors for each division of work. Question 1. Has been answered to our satisfaction within the first week since the effort began.

Question 2. Is also answered fairly quickly. We pick up the phone and call each sub with a very brief and pro-forma description of the scope of work and our roughly estimated dollar value of each division. Some calls go unanswered, or to numbers that are no longer in service. Some calls go straight through to someone in sales, or to a project manager, estimator, or company owner. They might come right out and tell you that they are booked for the year and cannot undertake the job. They might express an interest and have an immediate desire to meet with you at your earliest convenience. Another outcome may be that your call is referred to voice-mail and goes un-answered for a couple of days, and the call comes back from a guy working from his truck, who has not been in the office all week. It turns out that this guy is the sales department, project manager, estimator and owner all rolled up into one suit. It could be that he is having a bad week. It could be he is already in over his head. It could be that he is incapable of delegating anything. We don't know yet. He could be another Bob. He could be so focused on customer service that he did not make the time to call you right back because he was bending over backwards to help an existing customer. The reader is reminded that we are embarked on a process and do not judge a book by its cover. On our list of possible subs, we put a little asterisk by this sub and start taking notes of all of the phone calls. Hereinafter, let's call this new actor Jimmy.

We can write from experience and postulate the existence of a "natural wall" to growth. Among contractors, this occurs when annual revenues of about $10 million are reached. Until that tier is reached, will tend to see the job of estimator/project manager/superintendent all in the body of one, overworked person. When this point is reached the firm may roll back and be happy with this structure. If they wish to grow, these jobs must be split and delegated. This is a critical period of growth and many contractors can't pull it off. Once the split is made, they can go another $50-75 million before the nature of the business calls for another restructuring.

Our list of 15 prospects is culled to 10 or 12 based on the responses, or lack thereof to the initial phone call.

The third question is answered by way of a "Subcontractor Pre-Qualification Questionnaire" which we send to the remaining 12 prospects. We list the questions below and then discuss them category by category. The requirement for the general information should be self-explanatory.

General Information

    Name and Address of Contractor. Include satellite offices.

    URL of company homepage.

    Name and Phone Number of principal Point of Contact

    Number of Years in Business under the same name.

    Type of Business, Sole Proprietorship, Corporation, LLC, etc.

    Company works Union, Non-Union or both.

Questions about size and capability.

    Which division of work do you typically self-perform?

    Which divisions of work do you typically subcontract?

    Please state your annual revenues over the last five years.

    Please state the percent of these revenues generated by (1) Lump-Sum (2) Time and Material and (3) Cost and Fee contracts respectively.

    Please state the number of craft labor hours, (foremen and below) over the last five years.

    Number of management and administrative (office) personnel.

    Please provide an organization chart you would support for this project.

    How many square feet of office space do you occupy (Owned or Leased).

    How many square feet of shop space do you occupy (Owner or Leased).

Safety, bonds and Insurance

    Please state your experience modification rate for the last five years.

    Please attach a copy of your corporate safety plan.

    Do you employ a dedicated safety professional? Please provide his or her name and phone number.

    Please attach a copy of your insurance certificate.

    Please state the number of OSHA recordable incidents you have suffered in each of the last five years.

    Please state your approximate bonding capacity in millions.

    Have you ever defaulted on a bond?

References

    Provide the names and phone numbers of at least five customers you have contracted with over the last five years.

Miscellaneous

> Provide a list of owned or leased (not rented) capital equipment typically available.
>
> Provide resumes of management employees.

The questions as to *size and capability* are meant to give us a feeling for how the contractor will "fit" into the job. There is a natural force in the construction business that tends to drive contractors to bid more than they can handle. Construction is a tough business full of rough and tumble. There is seasonal variation to the opportunity set, hungry upstart competition, poorly developed bid packages and specifications, of course *NOT OURS!* Taken together with the temporary nature of construction, these competitive factors are not conducive to smooth sailing. The $10 million per year contractor I have so generally stereotyped in the preceding paragraph is more likely $5 million one year and $15 million the next. Jimmy might be dumpster diving for work one month and working his backlog 80 hours a week the next. They all want to grow and/or get to the next level. There is a tendency to overstate qualifications and capabilities. These questions are asked in an attempt to uncover and expose these issues. We do not want to hire a subcontractor who is in over his head from day one any more than one that is so huge that they will look at our job as a small one-off annoyance.

AXIOM 10

SIZE MATTERS, BUT BIGGER IS NOT ALWAYS BETTER

Beware the subcontractor that tries to present lower tier subs as his own forces. Contractually and legally this may be true, but from the point of view of management and control on the job site, this scheme fails. We need to be clear what these guys do with their own expertise and hands in the field. If a contractor represents that they do both mechanical and electrical work, they may be very well qualified for one division, and less so for the other. We do a site visit to inspect their investment in tools and discover they have lots of pipe wrenches and chain vises (for mechanical work) but have to rent a smart bender to shape electrical conduit. A smart bender is probably the first capital tool to be purchased by an electrical contractor. On the other hand, maybe they do all their electrical estimating in house, and own a half a dozen smart benders, but the mechanical capability comes from an acquisition they made six months ago, and half of those people fled for fear of change.

We ask the questions as to revenues and man-hours worked over the last five years, with an open mind. We hope to see a steadiness through a pragmatic, macroeconomic lens. From 2008 to 2014 we suffered a substantial recession, and most construction contractors did well to keep an even keel and manage losses. Otherwise, we would like to see steady growth, maybe one or two extraordinarily good or bad years that can be explained in the course of our investigation. Many contractors will engage in maintenance and/or service contracts that are bid once a year or less frequently, and the loss of such work can be difficult. Customers also change hands frequently and a well-qualified contractor can be on the outs due through no fault of their own just like that.

The question about man-hours gives us an idea as to how many "boots on the ground" our prospective subcontractor can typically support. A union contractor with a good cash position or access to credit can typically handle much greater volatility as to manpower and cash flow over the years compared with a smaller non-union contractor. What this means is that *all else being equal,* if we have a union and a non-union contractor and present them with a large (duration 1 year) job that might be worth 30% of their historical annual revenues, the relative stress on the non-union contractor will be greater than on the union contractor. If the job is 5% or 10% of the average revenues or man-hours, the difference in the relative stress on the contractors may be insignificant. If the job is more like a 50% spike to the norm we can probably expect problems with the non-union contractor.

We ask the question as to Lump-Sum vs. Time and Material work, because, as our project unfolds it will become desirable to use one or another of these contract vehicles to get the job done. Some contractors may do nearly 100% of their work on a lump-sum basis and asking them to do a job on T&M would cast them into the unknown. More on this subject will follow when we go to write the actual contract. For now, we just note the question as one more issue to be asked as we develop our understanding of the contractor.

Next, we need to get our hands around the administrative capabilities of our prospective contractors. Let's get back to Jimmy. Does he have adequate support or is he operating in a vacuum? Later there will be a chapter on documentation and control where we will discuss important documents and their administration:

The Contract
Drawings, sketches, their use and control
Request for Information (RFI)
Request for Proposal (RFP)
Submittals
Bulletins
Contract Letters
Letter of Transmittal
Field Orders
Change Orders
Time Sheets
Certified Payrolls
Requests for Progress Payments

Our contractor must have an administrative function capable of supporting the job at hand. At the core is the timely and smooth execution of "paperwork". Let's for the a moment stick to the old definition, whether the daily or weekly timesheet is filled out on some remote proxy server or filled in by hand with a pencil – if the document can be scanned or printed, stapled together or put in a binder, we are going to call it paperwork. The administrative requirements for the paperwork are that it is 1) complete, 2) orderly, and 3) instantly retrievable. Taken together, every one of the aforementioned documents carries a communication (instruction, question, answer, request etc.) a date, and forms a record of who was doing what, when, where, how, and why. For both contractor and owner, they form a bulwark against misunderstandings, disputes, delays, errors of both types, and expensive rework. When the job is complete to the satisfaction of all parties, these documents might be archived or discarded, but as the work progresses the maintenance and flow of the documents is crucial to progress, mutual understanding and profitability.

As to *safety, bonds and insurance*, if we are going to have five or six subs working for us in the course of the job, we want them all on the same page as to safety. All contractors are going to place a top priority on safety; owners too, but there is a divide between those that pay lip service to the subject and those of the new school who understand that better attitudes about safety can and will result in better productivity, morale and profitability. We ask each contractor his Experience Modification Rate over the last five years. The EMR is a number that serves as a multiplier on the contractor's workmen's compensation policy. The number is based primarily on the number and severity of workmen's comp claims made against the contractor, relative to the number of hours worked and other factors. It is usually computed as a three year moving average ending one year before rated year. For example, the modifier for 2014 would be the average of the modifier for the years 2011, 2012, and 2013. The median modifier by definition is 1.0. (half are better and half are worse). An unusually good one would be 0.5, and an unusually poor one would be 1.5.

For example, let's suppose our contractor kept 100 workers on fulltime payroll last year at a workmen's comp rate of $9.50 per hour.

The premium before modification is:

100 men * 2080 hours/man * $9.50 per hour = $1,976,000 per year

At a "good" EMR of 0.75 we pay $1,482,000
At a "bad" EMR of 1.25 we pay $2,470,000

And the difference between "good and bad" is $988,000

The annual reward or penalty for being better or worse than average is $494,000.

The point is that regardless of the size of the contractor, the EMR has a significant impact on the contractor's operating costs and profitability. At the end of the day these costs are passed along to us, the owner of the job.

There are a growing number of project owners, led for years by DuPont, that *require* at a minimum an EMR of .99 or less prior to any other consideration when qualifying their subcontractors. I adopted that policy in 1991, as an *ethical* requirement. How can one morally justify the use of labor that is *more likely than average* to be injured or cause harm to others?

It would also be irresponsible and problematic to ask a low EMR contractor to work side by side with a high EMR contractor in the same workspace. They will have fundamental differences as to priority. Work smart and safe vs. *"Git' er done."*

We ask for the GL policy certificate as part of our due diligence. The bonding question is asked for reference. There are many qualified contractors that never bid or worked a bonded job and some that do them all the time. In our case we will not require bonds and are not by law required to accept the lowest offer.

References are probably the most important qualifier and we shall thoroughly check them out. Our prospective contractor should be proud of his or her work and be more than willing to give us a tour of previous work. References are the only way we can get a feel for how the contractor might behave post award. If we have done a good job, there will be no surprises. Here is a shortlist of questions for the references.

- How did they get along with the owner and other subcontractors on a daily basis?
- How aggressively did they go after changes or dispute the scope of work.
- How was the quality of work?
- How well did they work within the ongoing operations
- Did they perform as promised?
- Did they meet the overall schedule and interim milestones?
- How were they with paperwork and invoicing?
- How did they do relative to your expectations?

If we have started the prequalification job correctly, we have a contractor that can meet the following basic requirements:

1. The contractor can supply the right amount of qualified and supervised LABOR as required.
2. The contractor will be able to meet all ADMINISTRATIVE requirements as to the timeliness and organization of all paperwork.
3. The contractor has the short term CASH or CREDIT or credit resources to meet a temporary and sharp increase in payroll.
4. The contractor is likely to get his work done SAFELY.
5. The contractor has a demonstrable track record of satisfactory PERFORMANCE.

Next, we prepare a brief cover letter for the questionnaire and send it out to our prospective subs:

David Glass

October 2015

Request for Qualifications

Dear Mr. Subcontractor,

A brief internet search and follow up phone call to Joe Smith indicated that your firm may be interested in a portion of our upcoming project which has a total estimated value of 10 million dollars. The job is scheduled to begin January 2016 and conclude by the end of December 2016. The exact schedule and value of each subcontract to be let is still under development, but we anticipate separate subcontracts for work associated with Mechanical, Civil/Concrete, Sheet Metal, and Rigging/Erection. Work will take place at Acme Manufacturing in __ County in __ State.

Prior to construction bidding or commercial negotiations, we are soliciting qualifications from prospective subcontractors. One immediate and non-negotiable requirement is that the contractor must have a current EMR of 1.00 or lower. If you do not meet this requirement, you shall be deemed unqualified for further consideration. If you meet the EMR requirement, please complete the attached questionnaire and return it along with any other information you feel would be helpful in our evaluation of your capabilities. The deadline for return receipt of the questionnaire is (two weeks) from the date of this letter.

If you have any questions please feel free to contact the undersigned. Contractors returning responsive questionnaires will be contacted for a site visit and subsequent discussions.

Our thanks in advance for your interest in this project. We appreciate your time and look forward to working with you as the project develops. Responses may be received by US mail or Adobe PDF at the choice of the subcontractor.

Sincerely,

D. Glass
Construction Manager

* * *

Now, two weeks have come and gone and the results might seem a little surprising. Of our 10 or 12 prospective subcontractors, by definition, the EMR requirement will have statistically eliminated half. At best, we are returned five or six questionnaires. Of the five or six, one will return a blank form and tell us to look at the video tour on their website, or peruse the enclosed brochure that only answers half of what we asked for. It might be a slick response, but between the lines, this sub has started out ignoring a simple request that would take an hour to prepare. Are they that lazy? Overconfident? Busy? Disobedient? Or has the tail already started trying to wag the dog? One or two guys don't even respond, and we just accept that for what it is. At this level of filtration, we don't care why. We have laid a substantial opportunity at the subs feet, and if they are too preoccupied with their existing backlog or bread and butter customers to even call with a reason why they cannot respond, it is just better to leave it there. We can lead a horse to water, but we can't make her drink. If they call after the deadline with an excuse and beg forgiveness, it is still best to cut bait. If they have that much trouble sorting the daily mail, setting a priority or meeting a simple deadline at such an early stage is a pretty fair indicator of future performance. Keep in mind that we gave them an hour's worth of work to do within a 2 week period of time.

Actions speak louder than words, in fact, *res ipsa loquitor.*

One or two subs will have completed the paperwork within 24 or 48 hours and will have called to ensure our receipt and express an interest in participating at the next level. All in all, of the 10 or 12 prospective subs we reached out to with our request for qualifications 3 or 4 have made the first cut.

The next step is a site visit. As some time has passed since we concluded the last conversation as to the progress of the engineering let's suppose we have our completed equipment list, a fairly complete P&ID drawing and a Layout drawing. We take what we got and go meet our subs. At this point in the selection process, further elimination is undesirable, unless we discover that the sub's references don't check out, or we uncover some misrepresentation between what they put on the questionnaire and what we see on the site visit.

We have some observations to make:

At the management / professional level:

Are we made to feel welcome or seen as an inconvenient intruder? The first impression will be a lasting one. Do we meet in a relatively neutral area like a conference room, or does the contractor sit behind his oversized desk in a cushy chair while you sit on some piece of plastic. Do they offer coffee and ask if you need to use the restroom. The meeting was for 10 am, why was your host preoccupied until 10:30? Do you just sit and wait or do they send someone else out immediately to get acquainted and show you around? These questions and impressions are important not so much as a check on professionalism and politeness, but at a more basic level, we are establishing our business relationship within a context where the *power* or business advantage is about to shift in favor of the contractor. If they act like they already have the power, be careful. Presently, we offer the contractor the prospect of a good and profitable job, and have a political advantage until the contract is inked, at which point we have placed the contractor in a higher position of power and depend on him or her to get the job done on target. If the contractor projects an air of superiority before the contract is in place, it is more likely that this air of superiority will turn into a stinking whirlwind of arrogance before the job is done. We hire contractors for their expertise, but at the same time we do not expect to be talked down to because they think they know everything. We are going to expect them to work in peace, harmony, and cooperation on the job site and at the craft and foreman level. If the upper management you are meeting with do not demonstrate a priority on cooperation with us, *the customer* – the one that still has the job to award, how do you think their foremen and superintendents are going to behave out in the field when they are running the show?

Within the office do we detect an air of cooperation between the cubicles? Are people confident with their heads up? Are the smiles and greetings genuine? Are they organized? Ask to see a copy of the form you sent them. Is it right at their fingertips in anticipation of a conversation, or is it all covered up somewhere even though they knew you were coming. Keep a mental score of how you are treated. You may be dealing with these people every day for the next year.

Now we take a walk around the shop. What we expect to see varies but we will observe a general correlation between the EMR and the orderliness of the shop area. They will have an outbound area where tools are sent to the field. Are they organized, clean, fit for use? Are the extension cords of industrial quality? Are there plenty of Ground Fault Interrupt pigtails? Are the blades and bits new and sharp? Is the floor broom swept? How is the lighting, noise and air quality? Are the hand tools in a good state of repair? Are the power tools numbered and is there a maintenance record for each one? Are things organized and on shelves. Do they prepare a manifest for each outbound and inbound shipment? Many of these questions may seem trivial, but a well-organized shop equates with a low overhead shop. Dull tools make bad cuts, slip and injure people, and drive EMRs higher. The culture of good housekeeping extends to the field: we do not want the craft servicing or looking around for tools, searching for the right drill bit. We want these things available for immediate and efficient use on our job. See if you can get a sense of the cooperation between the front office and the shop. They should be working in concert and cooperation, not in an atmosphere of animosity and isolation.

Finally, let's have a look at the capital equipment. There is not too much to be gleaned from a quick visit, but some good observations are available. Some contractors rely on leasing arrangements and maintain a relatively new fleet of equipment, others (most notably earthwork contractors) will keep equipment for 50 years and maintain it as long as parts are available. Essentially, the former will try to extract profits from a high fixed cost, low variable cost model, and the latter will try to extract profits from fully depreciated equipment carrying a higher variable cost. Our priority is that the equipment gets the work done in the field with satisfactory safe uptime. Lights, horns, brakes, reverse gear warning devices *must* work as new. Oil leaks from the powertrain or hydraulic systems *shall not be tolerated.* One of the more ridiculous things I have seen in the field was a contractor that installed plastic sheets and duct tape under leaky equipment, to serve as a diaper, only to have it catch and tear on an obstruction from time to time. Instead of having a continuous trail of drip, drip, drip leaking oil they would put up with a one gallon spill here or there. Again, here is your correlation between good organization, maintenance, and housekeeping with a lower EMR. Oil on concrete leads to slips, and falls. Falls lead to injury. Oil spills can lead to other undesirable consequences, such as unwanted fire.

I have attempted to present a brief primer on how to prequalify a potential subcontractor, and we now have 3 or 4 of each craft or division that will be invited to present a commercial offer for their services. While our contractors have met our requirements for prequalification, they remain untested and come to us with a diversity of attitudes. Let's return for a moment to that leaky equipment as an illustrative example.

One contractor has equipment that does not leak, another has equipment that leaks continuously, and a third has equipment that leaks all over when the silly plastic countermeasure fails. Again, we are buying an experience good, and at this point, unless we really nit-pic on the site visit, we don't know for sure what is behind door number one, door number two, or door number three. We want equipment that does not leak, and in this example, the probability is only 0.334 in our favor. There is also a .666 probability that a contractor will be wasting time and money cleaning things up, and an undetermined likelihood that one of these problematic actors will try to bill us for the plastic and duct tape. These are the kind of headaches that can be hard to identify prior to mobilization, but easy to neutralize using plain language in the bid documents. As we close this chapter, let's introduce the concept of "administrative controls".

We peremptorily remove the issue of the leaking equipment or remaining differences between the subs by introducing a simple requirement in the request for proposal. One of these "administrative controls" within a paragraph on equipment requirements reads as follows:

"Leaks of oil or other fluids from subcontractor's equipment shall not be tolerated. Work associated with leaking equipment shall be stopped, the leak cleaned up upon discovery, and the equipment removed from the site for evaluation and repair. Any costs associated with the stoppage and/or repair of equipment shall be at the sole expense of the subcontractor. Temporary catchments, pans or other means of containment are not acceptable. Equipment not immediately removed from service may be locked out at the sole discretion of the construction manager."

The foregoing language should serve as sufficient warning that you mean business. You also serve notice that you are the sheriff and final authority on the site. You have also established control over a potentially divisive issue. It is not required that you be the hard-ass and lock-out the offensive equipment; between the lines you have reserved the discretion to let the sub fix the leak on the spot if he can do the job on site without making more of a mess. There are many examples of administrative controls we have at our disposal to make the job go smoothly and to present each subcontractor with an identical set of expectations as to the shared and *general conditions* that we wish to put in place and enforce on the site. Additional examples will follow as we move to our next chapter and develop the commercial package and prepare the bid documents.

Let's imagine that we have 3 qualified subs from each of 5 divisions going forward:

- Electrical
- Sheet Metal
- Mechanical
- Millwrights
- Concrete and Civil

CHAPTER 5

Preparation of the Bid Documents Part 1 – Establishing the Mutual Expectations.

We foreshadowed this chapter with a brief statement as to a transfer of political power or leverage as we winnow our prospective subcontractors from the many to the few and finally the one for each craft or division of work to be done. In purely economic terms, when we first reach out to our subs, we have a monopoly on the perceived income our job may produce for the successful bidder. More simply, we have power over the sub in that we have money and the sub wants as much of it as he can get. This power is modulated in a positive or negative direction so far as the sub has access to other sources of income from other customers. If we move ahead with a project during a recession, we enjoy more power over the contractor than we would during periods of greater economic activity. When we execute the contract, we place our trust in that the contractor will perform as promised, but it remains that he or she now holds a virtual monopoly on the delivery of these construction products and services. Termination of a contract prior to completion (from the owner's point of view) of a job is never desirable, and we are guaranteed to lose time and money in any attempt to rebid and mobilize another contractor.

There is an old and still reliable school of school of thought whereby we, the buyer of these products and services would attempt to hold onto power through a variety of means such as liquidated damages for late delivery, payments which would lag to the extent possible any percentage of completion, and finally retention on the final payment that guarantees the contractor will reap no profit until the work is done to our final satisfaction. These rather blunt instruments do, in fact, work well. If we are the kind of buyer that just wants a relatively simple job done from a boilerplate specification, e.g.: big box, commercial, residential projects, we can just leave it at that, let the work go on a lump-sum basis, and check the site once a month. The job has been done so many times before that changes to the scope are unlikely unless we encounter something like unclassified excavation which requires a change order for shoring or blasting.

Our job is a different beast however. Time to completion is critical to attract the beloved early adopters. Our net present value analysis may be turned on its head if we come in second with the next-gen i-phone. Time may be so critical, in fact, that we bid the construction when the engineering is only 75% complete, and we anticipate, budget, and schedule the inevitable yet uncertain changes that will drive our job to completion. The question we answer here is, how do we structure the bid documents to control the cost of the coming changes in light of the fact that we must turn to our subcontractor, who has been placed in a position of higher relative, if not monopoly power.

There are essentially three contracting schemes and a couple of hybrid models that are traditionally used. As owner, each scheme presents us with a different degree of risk and cost, and different sharing of responsibilities between the owner and contractor. A brief discussion is in order:

1. Cost Plus Fixed Fee:

This scheme is good for highly complex and expensive (usually federal) projects where the quantities and outcomes are uncertain or unspecified. The subcontractor may have unique processes or proprietary know-how such that competition is not possible or limited. In this case, the owner has only sky-high or executive oversight. A goal is set and agreed to, and the work proceeds with the owner paying all costs, down to a ream of copy paper. Salaries and wages are certified and paid at cost. The contractor is given incentives and has broad latitude to prototype ideas to get the job done. The fee is guaranteed if the goal is met, but forfeit if the contractor misses the mark. I participated in such a project for a few years at the DOE Savannah River Site where the challenge was to remove millions of tons highly radioactive solid and non-pumpable sludge wastes from storage tanks and transport them to a plant where the waste was melted and mixed to a pourable viscosity and placed in stainless steel casks bound for permanent storage at Yucca Mountain, Nevada. The waste was such that it could not be approached by humans, and anything coming into contact with the waste would itself become a part of the disposal problem. This project has been going on for years with total costs in the hundreds of billions. Only the federal government has the resources to handle such a project and spread the risks across 300 million taxpayers. In this case, the initial goal was to produce 300 casks of more or less contained radioactivity per year. Any less than 300 and the contractor would recover costs, but no fee. Beyond 300, the contractor would profit handsomely, thereby providing incentives to develop technology leading to higher productivity. The contract is renegotiated every five years.

ASIDE

One of Vladimir Putin's political rivals, Alexander Litvinenko was poisoned with only fifty micrograms of polonium-210 in 2006. He died a miserable death three weeks after ingesting this miniscule amount in a cup of tea. The DOE complex manages thousands of tons of similar material without recordable incidents. In the course of discussions about fissionable materials and deadly isotopes with some of the older DuPont engineers who built the facility in the 1950's, they liked to say *"a little bit goes a long way"*

END OF ASIDE

2. Lump Sum

Another scheme, discussed previously, is that of a "lump sum" bid whereby the contractor is paid a fixed amount for a very specific scope of work. This scheme works best where the scope is well understood, there are clear battery limits, there is ample competition in the number and quality of qualified bidders, and significant changes are not anticipated. The size of the contract can be large or small, and the project can be simple or relatively complex. The only requirement is that the scope of work is well defined. A well specified $100 million prison facility bid with good competition might realistically return 6 bids with 2 or 3 percent between the high and low numbers. Under a lump sum contract, the contractor is in full control and responsible for how the work gets done, provided she meets schedule milestones and the final end date. The owner is placed in a relatively hands-off position, holding the contractor to spec and managing changes to the work if and when they arise. Changes may be administered by negotiating additional lump sum prices for extra work, and/or agreeing to a schedule of "time and materials" and/or "unit prices" for additional work, or a combination of all three depending on the circumstances. Imagine we have received bids for painting a yellow line down 10 miles of highway for $10,000. One thousand dollars per mile covers the paint, the operation of the truck, surface preparation, traffic control, temporary barricades, portable sanitation, etc. An extra mile on the same day on the same mobilization would reasonably cost $1,000 or 1000/5280 = $.19 per foot, if we had some paint left over…..

It would be unrealistic; however, to expect that contractor to buy another 55 gallon drum of paint (assuming that's the only way it comes), mobilize the same resources, come back next week and paint 25 feet at the same unit price. 25 x .19 = $4.75. On time and materials, the parties might reasonably agree to: truck, $100 per hour; escort car, $25 per hour; 2 drivers, 1 cone setter at $50 per hour each. Total labor $275 per hour plus a gallon of paint at $100. A 4 hour minimum (if they work that fast) costs you $1200. Now you appreciate the economies of scale and understand why you never see anyone painting the highway 25 feet at a time.

The downsides to lump sum arrangements are often painful and arise from a disparity in lack of experience on the part of the owner and years of experience on the part of the contractor. This comes in the form of high prices offered for lump-sum changes. Since the contractor is already there and working and has this *monopolistic* thing going, he will likely assert a territorial claim against work by others in the area under his control, or assert a claim for inefficiencies or delays if we try to get another contractor involved in his space. The control for changes increasing the scope of work can adequately be managed by releasing the work on the schedule of unit prices or T&M as discussed previously. On the other hand, when the scope is decreased, the contractor will try and hold on to his revenues by asserting that deleted materials have already been purchased, *(show me)* and there is a substantial restocking fee, *(show me)* and he has already incurred the overhead expenses associated with the deleted work in the form of some nebulous estimating or administrative effort. In the end you are offered $.50 on the dollar for the value of deleted work, and our contractor's margin tends to rise for doing less work. *Ouch! Dammit!* As owner, we do have some administrative controls at our disposal, which are often overlooked or neglected. We shall discuss this subject further going forward.

3. Time and Materials.

A third contracting scheme is to proceed the work entirely on the basis of time and materials, whereby we agree to hourly labor rates for the different classifications of labor, a percentage mark-up on materials purchased by the contractor, and a schedule of rates for rented equipment or equipment owned by the contractor. The negotiation and structure of these rates can form the subject of an entire subchapter, and detail is promised later. Time and material contracting is suitable for situations where the schedule is so compressed that we release parts of the work as the engineering design is available and the work is "shovel ready". As owner, we assume a greater degree of the construction risk, and must accept a more hands-on role. The contractor is virtually guaranteed a reasonable profit, and as owner, we are most flexible in response to engineering changes. We get overtime and can manage the manpower in the field as we see fit. One issue to overcome is that some contractors are so unaccustomed to working T&M that some of the administrative requirements might surprise them. For example, if we engage a large force on T&M we will probably require the services of an in-house or independent timekeeper who will keep track of the names and classifications of the workers on site, and for how many hours each day, what equipment was used, what materials were delivered and installed, etc. We will expect the contractor to present us with a corresponding *weekly* invoice for labor and materials. That is the only way we will be able to keep an accurate accounting of costs that will run to tens of thousands, or hundreds of thousands of dollars per day. We certify the invoice each week and pay the contractor promptly. She is here to work and get paid, not serve to finance the project.

ASIDE, on the subject of money

Let's keep in mind that our board of directors has approved the project and funding has been set aside to get the job done. Our controller, that guy or gal next to the corner office or out on the golf course, is partially compensated based on cash management, and might already have the project funds invested in money market or short term liquid commercial paper earning a small return each month. The controller has every incentive to delay payment to the contractor to the point where two big guys named Pauley and Pussy knock on the project office door with baseball bats looking for money.

This really happens......

It must be made clear from the start of the project within the internal organization that "due on receipt" means this afternoon or tomorrow, net 10 means net 10, and net 30 means net 30.

Remember in an earlier chapter how we had to deal with a relatively entrenched procurement department that was accustomed to the *routine?* The controller, also so comfortable with the routine, suddenly holds this big wad of *non-routine* corporate cash and the purse strings, and has this added incentive to hoard the greenbacks. Downstairs, and across the hall, far from the corner office, the contractor just wants to work and is not the Chase Manhattan Bank. Money is the *TOOL*, the *GREASE*, the *MEANS* and the ultimate resource that builds faith and gets the job done. We want the contractor to be part of the team and nothing we say or do speaks louder than delivering the first payment, by hand, when it is due and resolving billing errors on the spot. Keep things in perspective. If the sub is part of a $10 million job with funds already committed, and needs $20,000 upon mobilization, just give it to her. She has been through your qualification process, was responsive through the preconstruction period, and is working to add 20 millwrights to her payroll: *people* ready to work for you Monday who need to get paid and go buy groceries for their families on Friday. On your subcontractor's books, the 20 grand was already spent just bidding the job. It may take time, or not be possible, but the CM or PM should be in control of not only the scope and schedule, but also the cash flow of the project.

END OF ASIDE

Having discussed the basic schemes for bidding our contract, let's look at what we need to put together (at a minimum):

1. The scope of work, drawings, schedule, and technical requirements. – Includes equipment list, utility list, device list, etc.
2. Specifications and List of Submittals
3. A complete schedule of values with a solid work breakdown structure.
4. General conditions

1. Developing the Scope of Work.

The scope of work is broken down into our expectations as to what we require from each craft or division. It can be a relatively simple statement, for example: (All work as shown on the drawings, as specified)….. The millwrights will be charged with receiving, rigging, setting, and assembly of the equipment as shown on our layout drawing. In order to bid this job, the rigging contractor has to know how big and heavy each lift is going to be. We might be lifting small cases of machinery, several from an over the road flatbed, or pulled from an intermodal shipping container. The machinery can also be relatively fragile, and more or less subject to damage. For more complicated lifts, we will require a submittal called a rigging plan that will describe the lifting method, the location of lift points (developed in coordination with the machinery maker) the rigging gear to be used, and trigonometric calculations of the loads on each and every shackle and sling. The rigging plan should be certified by a professional engineer and take into account the allowable wind and weather conditions on the day of the lift. It should also be checked for soil conditions as to yield strength and moisture content. Special provisions may be necessary to spread the ground loads. Getting the forklift stuck on a soft piece of ground is one type of problem; toppling a crane is usually a type two error.

2. General notes on the specification of work / techniques / materials:

Another special consideration for assembly and other mechanical connections that is often overlooked is a torque and fit up specification for the nuts and bolts. Some come down to a preference, or a not very scientific procedure: for the assembly of towers for the FAA, the structural hardware consisted of a nut, bolt, and lock-washer, which was made up to hand tight, marked with a sharpie, and then tightened at least 1/3 of a rotation, or 120 degrees as measured by a visual inspection. A better specified requirement for a hazardous materials or, critical process pipe, might require flanges to be made up with a bolt, lock, and flat washer of a certain ASME *grade* certified by the manufacturer, with a specified locking compound or lubricant, and tightened in a certain sequence, to a specified torque. To verify that this job gets done right, we might also be required to employ the services of an independent construction observer or inspector. With the concrete work, we will specify a certain particle size distribution on the aggregate, a certain allowable initial slump, a specified 7 and 30 day compressive strength, and a certain number of test cylinders per load or so many cubic yards if we are doing a continuous pour. There are other factors to be considered. For example, concrete can be mixed with different chemicals to provide high early strength, or retard the set and improve the viscosity to allow better pumpability which is important in oilfield or high rise applications.

Moving on to our electrical specifications, we will have different requirements as prescribed by the National Electric Code (NEC) for wire sizes, types of conductors, requirements for splices, clear space in front of panels, and the number of conductors that can go into a single run of conduit. Between the requirements of the NEC, The National Electric Manufacturers Association (NEMA), and The National Fire Protection Association (NFPA) almost every electrical requirement is spelled out for us in acceptable detail. Owners may have additional requirements based on prior experience and lessons learned, such as the separation or shielding between power and control wiring. PLC's might be Allen-Bradley, Toshiba, Siemens or Omron; they might all do the same thing but run different software. One type of PLC might be desirable due to communications and software requirements of the existing plant. We might require stainless steel, coated conduit, EMT, or rigid conduit based on some proprietary process knowledge.

On the mechanical side, the ductwork will come in different standard sizes, and thickness (gauge) of material. There is ongoing confusion on the subject of thickness because a given gauge number returns different thicknesses for different types of material. The disparity is best understood when the information is reduced to a Table 1:

Standard sheet metal gauges						
Gauge	U.S. standard for sheet and plate iron and steel decimal inch (mm)	Steel inch (mm)	Galvanized steel inch (mm)	Stainless steel inch (mm)	Aluminium inch (mm)	Zinc inch (mm)
3	0.2500 (6.35)	0.2391 (6.07)	......	......	......	0.006 (0.15)
4	0.2344 (5.95)	0.2242 (5.69)	......	......	......	0.008 (0.20)
5	0.2188 (5.56)	0.2092 (5.31)	......	......	......	0.010 (0.25)
6	0.2031 (5.16)	0.1943 (4.94)	......	......	0.162 (4.1)	0.012 (0.30)
7	0.1875 (4.76)	0.1793 (4.55)	......	0.1875 (4.76)	0.1443 (3.67)	0.014 (0.36)
8	0.1719 (4.37)	0.1644 (4.18)	0.1681 (4.27)	0.1719 (4.37)	0.1285 (3.26)	0.016 (0.41)
9	0.1563 (3.97)	0.1495 (3.80)	0.1532 (3.89)	0.1563 (3.97)	0.1144 (2.91)	0.018 (0.46)
10	0.1406 (3.57)	0.1345 (3.42)	0.1382 (3.51)	0.1406 (3.57)	0.1019 (2.59)	0.020 (0.51)
11	0.1250 (3.18)	0.1196 (3.04)	0.1233 (3.13)	0.1250 (3.18)	0.0907 (2.30)	0.024 (0.61)
12	0.1094 (2.78)	0.1046 (2.66)	0.1084 (2.75)	0.1094 (2.78)	0.0808 (2.05)	0.028 (0.71)
13	0.0938 (2.38)	0.0897 (2.28)	0.0934 (2.37)	0.094 (2.4)	0.072 (1.8)	0.032 (0.81)
14	0.0781 (1.98)	0.0747 (1.90)	0.0785 (1.99)	0.0781 (1.98)	0.0641 (1.63)	0.036 (0.91)
15	0.0703 (1.79)	0.0673 (1.71)	0.0710 (1.80)	0.07 (1.8)	0.057 (1.4)	0.040 (1.0)
16	0.0625 (1.59)	0.0598 (1.52)	0.0635 (1.61)	0.0595 (1.51)	0.0508 (1.29)	0.045 (1.1)
17	0.0563 (1.43)	0.0538 (1.37)	0.0575 (1.46)	0.056 (1.4)	0.045 (1.1)	0.050 (1.3)
18	0.0500 (1.27)	0.0478 (1.21)	0.0516 (1.31)	0.0500 (1.27)	0.0403 (1.02)	0.055 (1.4)
19	0.0438 (1.11)	0.0418 (1.06)	0.0456 (1.16)	0.044 (1.1)	0.036 (0.91)	0.060 (1.5)
20	0.0375 (0.95)	0.0359 (0.91)	0.0396 (1.01)	0.0375 (0.95)	0.0320 (0.81)	0.070 (1.8)
21	0.0344 (0.87)	0.0329 (0.84)	0.0366 (0.93)	0.034 (0.86)	0.028 (0.71)	0.080 (2.0)
22	0.0313 (0.80)	0.0299 (0.76)	0.0336 (0.85)	0.031 (0.79)	0.025 (0.64)	0.090 (2.3)
23	0.0281 (0.71)	0.0269 (0.68)	0.0306 (0.78)	0.028 (0.71)	0.023 (0.58)	0.100 (2.5)
24	0.0250 (0.64)	0.0239 (0.61)	0.0276 (0.70)	0.025 (0.64)	0.02 (0.51)	0.125 (3.2)

I present this partial table for information only, (gauge numbers run up to 36, and material having a gauge number less than three is referred to as plate). We also will not go into a discussion on metallurgy tolerances, or performance. As a project manager, what you need to know is that the higher the gauge number, the thinner the material, the thinner the material, the less rigidity and the more susceptible to the effects of corrosion. It should also go without saying, the higher the gauge number, the less mass and cost per unit of area.

A good sheet metal design, as with any other suitable design, will balance service life, cost of installation, cost of materials, environmental factors, and on and on. A poor design will come to you at a high cost per foot with the wrong thickness and materials.

Finally, we will close this section on our specifications by noting that the same type of situation exists for piping materials. There are different wall thicknesses associated with the same nominal sizes pipe, commonly referred to as schedule 10, 20, 40, 80 and 160 materials.

Nominal Pipe Size	Outside Diameter	Schedule												
		10	20	30	STD	40	60	XS	80	100	120	140	160	XXS
(in)	(in)						Wall Thickness (in)							
1/8	0.405				0.068	0.068		0.095	0.095					
1/4	0.54				0.088	0.088		0.119	0.119					
3/8	0.675				0.091	0.091		0.126	0.126					
1/2	0.84				0.109	0.109		0.147	0.147				0.188	0.294
3/4	1.05				0.113	0.113		0.154	0.154				0.219	0.308
1	1.315				0.133	0.133		0.179	0.179				0.25	0.358
1 1/4	1.66				0.14	0.14		0.191	0.191				0.25	0.382
1 1/2	1.9				0.145	0.145		0.2	0.2				0.281	0.4
2	2.375				0.154	0.154		0.218	0.218				0.344	0.436
2 1/2	2.875				0.203	0.203		0.276	0.276				0.375	0.552
3	3.5				0.216	0.216		0.3	0.3				0.438	0.6
3 1/2	4				0.226	0.226		0.318	0.318					
4	4.5				0.237	0.237		0.337	0.337		0.438		0.531	0.674
5	5.563				0.258	0.258		0.375	0.375		0.5		0.625	0.75
6	6.625				0.28	0.28		0.432	0.432		0.562		0.719	0.864
8	8.625		0.25	0.277	0.322	0.322	0.406	0.5	0.5	0.594	0.719	0.812	0.906	0.875
10	10.75		0.25	0.307	0.365	0.365	0.5	0.5	0.594	0.719	0.844	1	1.125	1
12	12.75		0.25	0.33	0.375	0.406	0.562	0.5	0.688	0.844	1	1.125	1.312	1
14	14	0.25	0.312	0.375	0.375	0.438	0.594	0.5	0.75	0.938	1.094	1.25	1.406	
16	16	0.25	0.312	0.375	0.375	0.5	0.656	0.5	0.844	1.031	1.219	1.438	1.594	
18	18	0.25	0.312	0.438	0.375	0.562	0.75	0.5	0.938	1.156	1.375	1.562	1.781	
20	20	0.25	0.375	0.5	0.375	0.594	0.812	0.5	1.031	1.28	1.5	1.75	1.968	
22	22	0.25	0.375	0.5	0.375		0.875	0.5	1.125	1.375	1.625	1.875	2.125	
24	24	0.25	0.375	0.562	0.375	0.688	0.989	0.5	1.219	1.531	1.812	2.062	2.344	
30	30	0.312	0.5	0.625	0.375			0.5						
32	32	0.312	0.5	0.625	0.375	0.688								
34	34	0.312	0.5	0.625	0.375	0.688								
36	36	0.312	0.5	0.625	0.375	0.75								
42	42		0.5	0.625	0.375	0.75								

STD - Standard
XS - Extra Strong
XXS - Double Extra Strong

The greater the schedule number, the greater the thickness, pressure rating, and cost per foot of the same nominally sized pipe. There are also a very wide variety of materials available, any number of plastics, steels, or specialty metals are in use depending on many more design considerations. There are many different specifications as to connections – flanges, threads, victolic, compression, socket weld or butt weld, welding technique (MIG, TIG, Stick, etc.), and the quality control on welds from a simple visual inspection and pressure test, up to 100% x-ray examination.

What I mean to convey to the reader in the foregoing discussion, is that a good specification, a product of the engineering service, will completely cover the *quality* of our installation, the *exact material* used, and which division or subcontractor will be responsible for doing the work. As owner, you must know that overkill on the specification can easily double the overall cost of the work, while poorly specified materials and requirements on labor will lead to unsatisfactory work as well as bids which contain unspecified assumptions resulting in minimum quality and big change orders and headaches later. Refer, once again to our second axiom. Part of the planning means that we ask a lot of questions during the development of the specifications. We scour the specifications for constructability and value. Some consulting engineers will have been at it so long that they issue the same drawings for piping supports for almost all applications, and we wind up with twice as many as required. On the other hand, an in-house engineer can be so mismanaged and under such financial pressure that other types of supports (- like centralizers in oilfield applications) are deleted and lead to big problems such as the Deepwater Horizon.

During the construction phase, there are also some oddball circumstances where the specifications do not match agreements among the unions as to who gets to do what. For example, a pipefitter's local might claim jurisdiction over the installation of all valves on a job, even though one of them might weigh 100 tons, enough to require a engineered rigging plan. Then again, a Kentucky millwright (*qualified to lift anything*) once complained to me that he was just not allowed to install a solenoid valve on an instrument because it had some *"waars"* aka wires sticking out of it and for lack of a local rule, was, by his own theory, electrical work. The electricians claimed extra money for any work on a solenoid other than the wire terminations, and when the pipefitters heard about the issue, they claimed the only scope they had was to connect the pneumatic tubing to the valve actuator.

The easiest way to get it straight is to ask the business agent who does what at this level, or ask the contractors during your bid meeting and issue a clarification by way of a bid addenda, before the bids are received.

Finally, I'll leave the subject of specifications by noting that for as many different materials and equipment there are available, there is a trade association, engineer's society or some government regulator having rules or a boilerplate spec for the work. Incorporation of these specifications (as applicable) in our bid package form a very effective administrative control to keep the contractors on the same set of expectations throughout the bid process. There are also differences between the states in specifications for public works. The best example is every state having its roads built to a certain standard called out by the Department of Transportation – (Penndot, NJDOT, Caltrans, etc.).

Our administrative control simply requires the contractor to work "in accordance with", or IAW one or more of the standards. Another way to do it and save lots of paper is by stating "the following are incorporated by reference." Here's a quick glossary of the more widely cited standards and organizations:

NEC	National Electric Code
NEMA	National Electric Manufacturers Association
SAE	Society of Automotive Engineers
SPE	Society of Petroleum Enginneers
ASME	American Society of Mechanical Engineers
ASCE	American Society of Civil (and/or) Chemical Engineers
AIA	American Institute of Architects
ASHRAE	American Society of Heating, Refrigeration and Air-Conditioning Engineers
AASHTO	American Association of State Highway Transportation Officials.
SMACNA	Sheet Metal and Air Conditioning Contractor's National Association.
API	American Petroleum Institute
STI	Steel Tank Institute
ACI	American Concrete Institute
ASTM	Formerly known as the American Society for Testing and Materials, now known as ASTM International
ANSI	American National Standards Institute
NFPA	National Fire Protection Association
Federal Codes:	
OSHA	Occupational Safety and Health Administration - regulates conditions for construction and operating plants.

DOE	Department of Energy – special requirements for nuclear plants and operations.
DOD	Special Requirements for Department of Defense Contracts
FAR	Federal Acquisition Regulations

3. Building the Schedule of Values:

Now that we are nearly complete with parts one and two of our bid package, we will jump ahead a little to the formulation of our *schedule of values*, because it is the schedule of values that form our basis for interpreting and evaluating the bids, as well as the backbone of our man loaded construction schedule. In the old days, the contractor was usually permitted to submit an *unspecified* schedule of values as a kind of free form statement of what things would cost the owner along the way. The unwitting owner might allow a line item for "Mobilization" which would be intended to cover the cost of getting the job started, and the contractor would fill that line in up to the maximum value permitted by the owner – "frontloading" the job to the extent possible. This practice would mean the contractor gets as much cash as possible up front, and unwittingly, we give him a double edged sword to beat us with later, because the subsequent line items actually costing much more in time and effort carry less contractual value and we are able to extract less if there are deletions in the scope of work. If we allow frontloading, we are just leaving the contractor with a growing monopoly on the remaining effort.

It is much better to be forthright and honest with each other at all times. If he *needs* the cash up front, as we discussed earlier, it is better to agree to pay promptly for materials delivered to the site or having been documented as delivered to the contractors yard. Go ahead and give him his cost and mark-up a little early on delivered materials net 7 or 10, and he or she can be on their own to work net 30, or longer, as agreed, with their suppliers. Doing things this way allow us to maintain a security interest in the materials and keeps our schedule of values more or less intact and in keeping with what is going on with the work.

Further, as we build the schedule of values, we maintain the completeness of each line item by including columns of costs reflecting the relative cost of the associated overhead, profit, materials, and equipment. Thereby maintaining a schedule that can stand in its own line by line.

Here is what I mean as an example of some mechanical work for a project completed with a good result in 2014.

Process Utilities Scope of Work -Bid Form

Line Item	Area	Drawing	Scope	Reference Documents	Total Linear Feet	Labor Hours	Labor $	Materials $	Equipment & Consumables $	OH&P $	Total $
1	TA 1 and 2	8	6", 4"and 1-1/4" from Pump Room wall near D9 to TA 1&2 plus 16 drops. Continue header to CL 4.		1297	1743	$102,316	$29,269	$4,400	$17,200	$153,185
2	TA 1 and 2	M1.11	Insulation for Glycol Piping above		807	198	$17,480	$6,927	$1,200	$640	$26,247
3	FCU1	M1.12	Continuation of 1-1/4" from CL4 to FCU in MCC Room		125	115	$8,200	$616	$1,300	$1,500	$11,856
4	FCU1	M1.12	Insulation for Glycol Piping above		186	65	$5,880	$1,658	$560	$450	$8,548
5	FCU1	M1.12	Installation of Trane FCU and Ductwork in MCC room	FCU 1 15 2013 (MS Word), M1.15	NA	64	$4,352	$1,140	$450	$325	$6,267
6	Pump Room	M1.13	Receive and set P1-P8, Chiller, Glycol Tank, Tower Water Tank, and HX1	LiDestri Foods Process Chillers (MS Word), Pump HX submittal (pdf)	NA	84	$5,712	$4,337	$575	$356	$10,980
7	Pump Room	M1.13	All piping in Pump Room, and Between Room and Chiller. **Submit isometric for approval.**		427	2134	$120,682	$66,624	$3,825	$23,144	$214,275
8	Pump Room	M1.13	Insulation for above	Marley PF501130 P,M,S,G	255	134	$11,760	$3,934	$553	$296	$16,543
9	Roof	M1.14	Receive and set Cooling tower on existing support steel. **Submit transition piece drawing for approval,** fabricate and install transition. Complete tower piping not otherwise described above		60	110	$5,440	$5,700	$10,951	$3,900	$10,390
10	TA 1 and 2	P2.11	Compressed air tie-in at B6 to TA 1&2 and across rail shed	P5.01, P7.01, P0.01	720	449	$25,390	$6,353	$2,150	$5,100	$38,993
11	Filling Area	P2.11	Compressed air from tie-in at B4 B-F, 2-3.5 (Drops as indicated)	P5.01, P7.01, P0.01	1080	340	$22,508	$13,129	$2,150	$4,050	$41,837
12	Mix-Blend, CIP	P2.11	Remaining CA from tie-ins near B4, C5, and D5 (presumed) to Mix-Blend, CIP and drum stations	P5.01, P7.01, P0.01	710	479	$25,896	$9,466	$4,150	$5,500	$45,012
13	Filling Area	P2.11	City water from tie-in at B2.5, to filling area (Drops as indicated)	P5.01, P7.01, P0.01	925	376	$22,759	$20,293	$2,100	$4,350	$49,502
14	Mix-Blend, Almix, CIP	P2.11	Remaining CW from tie-ins at B3A, 4D(2) to Mix-Blend, Almix and CIP	P5.01, P7.01, P0.01	740	655	$35,629	$23,703	$4,100	$7,050	$70,482
15	Mix Blend	P2.11	Install Nano water, complete as indicated.	P5.01, P7.01, P0.01	376	268	$15,756	$57,415	$2,250	$4,200	$79,621
16	First Floor (all)	M2.11	Install all steam and condensate piping as shown	M5.01, M0.01, M7.01	1200	853	$50,746	$41,916	$2,250	$8,600	$103,512
17	First Floor (all)	M2.11	Insulation for above	M5.01, M0.01, M7.01	2300	1120	$81,564	$14,235	$3,227	$1,457	$100,483
18	First Floor (all)	M5.01	Installation of remaining steam / condensate stations, traps, etc.	M0.01, M2.11, M701	NA						$0
				Totals:	11208	9187	$562,070	$306,715	$46,191	$88,118	$987,733

106

The left side of this form has been developed by the project manager or project engineer, and we have pared down the scope of mechanical work to 18 manageable segments or sub-scopes. The basic idea in the development of this document is to take-off a discreet area, such as a room, machine station, or continuous length of pipe that can be accurately measured for quantity on the drawings and verified in the field.

For the beginner, we must digress for a moment: Every plant layout drawing will be placed on a grid in accordance with column lines, letters on one axis and numbers on the other. In addition, there will be a "plant North" assigned for reference. Taken together, we have a true map of our facility. Care must be taken to assure that this grid is universal and maintained as accurate across all drawings and vendors. We do not want our engineering company to have column A1 in the far Southwest and our main equipment vendor to be working on a grid with A1 in the far Northeast. Further, vertical elevations must also be referenced to a universal benchmark. This might be as simple as picking a spot on the floor and calling it 0.0, or 100 if you like. Sometimes the elevation will be established from the very beginning of the job with an elevation shot from a nearby United States Geologic Survey Benchmark. Again, the elevations must be universal across all designers, vendors and contractors. While the foregoing may seem trite, I once had a role on a job where the general contractor had placed a scribe on a building column to mark an elevation of 1206 feet above mean sea level, and all of the architectural drawings were referenced to MSL. Months later, after the GC had demobilized, the scribe remained, and it just so happened that the 1206 mark was about 1306 millimeters above the finished floor. It took a lengthy meeting with the Japanese vendors, who were in fact working with millimeters to explain the mark referenced feet above MSL, when it appeared on the column to be wrong by 100 mm or about six inches. In fact, the matter was resolved only after some equipment had been fabricated by different vendors and the elevations at the battery limit between them had to be resolved with a torch. It is the job of the CM or PM to maintain a master layout as the project develops and make sure the vendors are kept up to date on the same and do not pile equipment on one another or disregard locations reserved by a platform vendor for locating mezzanine columns.

Here is an example of a drawing with a suitable grid, column F8 in the far Northeast.

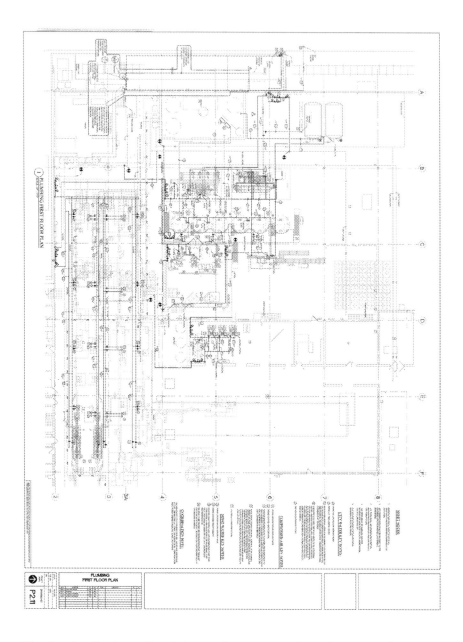

The South of column line 2 is not shown or used on this project because it was just an office area outside the scope of work. A suitable grid is critical to obtaining uniform estimates from our subcontractors and saves time in asking and answering questions. In our schedule of values, Line item 11 refers to all of the compressed air piping shown on the drawings within the area bounded by columns B2, F2, and another line we draw between B3.5 and F3.5. A full size drawing of P2.11 referenced above would show 3 runs of 3" copper pipe form B to F, at an elevation 20 feet above the floor, the number of valves, and the location of the service drops to each piece of equipment. The equipment drawings will show the battery limit and elevation of the service drops, and other reference drawings will specify the manufacturer of the valves and a typical support and spacing detail. At this level of detail, senior estimators should come up with the exact bill of materials and length of pipe to the nearest 20 foot joint. If time allows, our project engineer does his own estimate on the pipe and comes up with 1120 feet compared with our subcontractor's estimate of 1080, and we have a level of comfort that the sub has done his estimate correctly. If we receive other bids of 500 or 1500 feet, we have basis for questions. Although, in the end, if we want the job done on a lump-sum of $987,733, we want our sub to have a perfect understanding of his scope. Maybe one guy only saw two runs of pipe and the other gal saw four when the correct number was three. If the bids come in tightly packed, we do not want to eliminate a sub because they had a misunderstanding on a material take-off. As to the spreadsheet columns on our schedule of values for labor hours and labor dollars, we expect our bidders to show a greater degree of variation. One may be more conservative than the other on the hours required, another may be deliberately packing overhead and profit into the labor and making it seem that the OH&P column is slim. In any event, these SOV columns give us a good basis for comparison. The percentage difference between the columns on Labor Cost, Hours, and OH&P should show some consistency between the bids.

One additional point on material take-offs: Years ago, drawings were done by hand by people sitting in big rooms at tables with pencils, stencils, rapidographs, scales, and erasers. They would draw on translucent paper or other material known as mylar or modern day vellum. The finished sheets would then be fed through an ammonia based process and the image could be copied to opaque paper as the image transfer maintained the exact scale of the original. Architects used scales with fractions of inches representing one foot and engineers would use scales having 10, 20, 30, 40, 50, or 60 divisions to the inch. The point is that the "blueprints" resulting from the copy process carried the same scale as the original. When doing a take-off, the estimator would just take the appropriate scale and take the unit lengths straight off the drawing. Things changed as technology developed, the use of PCs increased, and the price of printers and plotters dropped. There is a greater variety of paper and software in use. Architects and engineers still use physical scales when doing design work in auto-cad and micro-station, and drawings are still produced electronically to exact scale, but anyone making a copy can inadvertently change the scale on the duplicates. Drawings are more likely to be transmitted in an adobe Pdf format, and 11" x 17" (ANSI B size), 22"x34" (ANSI D) are nearly standard in the US. The use of ISO sizes is more common in the rest of the world. Since you are specifying everything, you might even go ahead and specify which size paper *shall* be used.

A novice or inexperienced engineer might be confused when the same drawing's title block says ¼" = 1 foot, and there is a note elsewhere on the same sheet of paper that says "NOT TO SCALE" or "DO NOT SCALE". What is meant is that the scale in the title block more than likely is no longer accurate. You may have been handed an exact 50% reduction, in which case 1/8" = 1 foot, however it is much more likely that what you hold in your hand has been reduced or magnified to any random degree. This leads us to our:

AXIOM 11

THE CONTRACTOR SHALL VERIFY ALL DIMENSIONS, AND WE MEAN EACH AND EVERY LAST ONE OF THEM

We start with our layout drawing, which will show some distance between the building columns, for example:

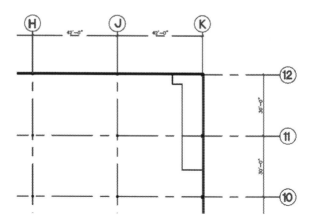

This Northeast corner of our building shows that we have a north and south dimension of 30 feet between the columns, and an east west dimension of 42 feet. We verify these dimensions by sending two guys out to the field and measure the actual distance between the columns. A carpenter or millwright would know that it matters not if we measure north edge to north edge or east edge to east edge as long as we are consistent; an engineer would make it complicated by guessing or measuring to the center of the columns and then measuring between them. We send two guys out because of

AXIOM 12

MEASURE TWICE AND CUT ONCE

They *independently* measure the distances, north to south and east to west, without having seen the drawing, or being told what to expect, and return the distances having written them down on separate paper and without discussion. If we are lucky, they both return the same result, which is 29'-11$^{1/2}$" and 42'-$^{1/8}$" while the measured distances do not exactly match the corresponding dimensions on the drawings, they are close enough to verify this particular piece of civil work. At this point, we can say we have verified the dimensions on the drawing as 30' and 42' between the columns, and are certain there is not a systemic typo on the printed page. We might want to measure a few more places for more certainty. Despite checking, some errors can go undetected on drawings for several revisions.

The next thing we do is verify the scale in the title block on the drawing. We take our engineers scale and lay it across the 42' lines on the drawing and come up with 42' = 2.70" that *ratio* is the correct scale for the East and West lines on the drawing. Now, take 30/42 x 2.70 = 1.93" and then measure between the 30' lines on the drawing. We have every expectation that the North-South measurement is exactly 1.93 and the *ratio holds across the paper*. We have verified dimensions and *derived* or verified the scale on the drawings. I have been around the block more than once and have never received a drawing having different NS-EW scales, but truth be told I have had occasion to reject them. Hence the rather anal foregoing discussion on measurement. Keep in mind how easy it can be to stretch things out of ratio with a picture file in Windows, and how much a per-foot creeping error might cost on a bridge or tunnel. At the end of the cold war, on the subject of nuclear arms reduction,

Ronald Reagan said "Trust but Verify."

When working with units and measures,

Your Humble Author says: - "trust nothing, verify everything."

Finally, we complete the discussion on the schedule of values by noting that we will add some language to the contract which states something like:

"Contractor agrees that the hours, dollars, linear units, and derivatives thereof contained in the schedule of values shall be used as the basis of calculation for any addition or deletion to the scope of work contained within each line item."

At first, the contractor will not like this language because he has rarely had it incorporated into a contract, but as owner or owner's rep, we have maintained our fiduciary duty to our shareholders by keeping a level playing field as to the value of additions or deletions to the scope. Second, as to an element of human behavior, we begin to gain the respect of the contractor by showing a certain *savoir faire*. He knows that we have a genuine interest in fairness, and at the same time realizes we are hip to his pathways to nefarious behavior. He also knows that we have spent a substantial amount understanding our own drawings. How else could we have come up with such a perfect schedule of values?

At this point in the development of our bid package, I assert that we have covered the fundamental requirements. Our engineering group has produced drawings and specifications that tell the contractor what work is be done, using specific types of materials and techniques. We have enough detail that the contractor can estimate and measure the quantities of materials and labor required to do a complete job. At this point, the contractor's understanding is at 85%. The next 10% has to do with what we call general conditions, a discussion of which follows below.

4. Understanding the General Conditions.

"General Conditions" are often misunderstood, poorly defined, unbudgeted, unspecified, and confusing. Sometimes even referred to as general requirements or overhead. These are the things more or less taken for granted in everyday office life, but unplanned for and unusual expenses when we move to construction. Our bid package must contain language as to the general conditions, including special plant requirements, especially when we have multiple contractors sharing them.

If we are working within an existing plant, the general conditions are well established and taken for granted. The mobilization of the contractors will cause a major disruption. If we are on a Greenfield site, the general conditions need to be established. We establish the shared norms of behavior among the contractors. Uniform management of the general conditions is crucial to the overall *happiness* on the site.

We will define General Conditions as: "overarching shared and common conditions, requirements, limitations, and expectations, present on the site and incorporated within the various contracts whether specifically applicable to each contractor or not."

It is easiest to make a list:

Traffic Control and Parking
Temporary offices and trailers
Temporary Roads or other Construction
Plant Access
Working Hours
Designated work areas, aisle-ways, laydowns, equipment parking
Receiving
Security
Maintenance of Roads
Sanitation and Water
Use of Plant Facilities
Breaks and Lunch
Tobacco Use
Housekeeping
Disposal
Temporary Power
Snow removal

In the foregoing paragraphs, we have gone to considerable length to specify exactly, the materials, techniques, methods etc. required to perform the scope of work to a set of mutual expectations. Until now, we have not made a statement as to the management or allocation of costs associated with these general conditions, which are common and shared on the site. Unless otherwise noted, the contractor probably has not included, and certainly will assert that he does not have any of these costs in his bid. This is where the owner's representative, or, on a larger job, the construction manager takes the lead.

Earlier, I asserted that we might set-aside 10% of our costs for the general conditions. Assuming that none are included in our above bid, we make a budget of $99,000. Further, the contractor has bid of 9187 hours. For now, let's suppose we will do the work with an average of 12 workers. For purely illustrative purposes, the duration of the job will be 12m x 40hrs/week =480 hrs/week…..9187hrs/480hrs/week = 19 weeks. Details on man-loading schedules and start to finish durations is another chapter.

Let's continue point by point, on the simple assumption that we go 19 weeks on site. The duration is important because time is money, and general conditions are supported by things we rent and cost money to maintain.

Traffic and Parking

First, we agree with the plant as to parking. Perhaps the existing lot is large enough, but we need to allocate specific spaces for the craft labor so as to keep it distinct and separate from the plant employees. Next, we will need a designated space for construction trailers with easy access to the work area, and with designated parking for the trailer personnel. Construction trailers should be close together to facilitate communications among the contractors. For temporary power to the trailers, if we did not heretofore specify that each contractor was responsible for his own, we did make a budget for the electrical contractor to provide a common meter stand with one feed from the transformer on the pole to 5 or 6 meter sockets, or maybe one meter for all the trailers, and we foot the whole bill. The important thing is that we made a plan and a budget.

Ingress and Egress

Next, we need a plan for plant access and egress. There is usually some security system in place and if our plant has little turnover, it might rely on facial recognition by the guard, a prox card system, a palm reader, or other device. Our job is about to cause a major disruption to the system and again we need to make a plan. Perhaps we designate a contractor's entrance, or issue cards to everybody or just to the foremen, or one per contractor. The schemes can go on and on, but the first thing the sub will do is assert a claim for inefficiency if it takes any more than a few minutes to get in and out of the plant. We get the understanding up front and put the delay in the bid if there is no easy way to get the doors wide open. At DOE Savannah River, there would be an end of day whistle when the work would stop, and another 15 minutes later when people were allowed to start for the gate. They'd cue up like Holsteins and leave through a common turnstile. Some plants want everyone in by a certain time and out eight hours later. Chances are, the plant schedule doesn't meet with the satisfaction of all the subs. Another reason for meetings and prior agreement.

Laydowns and Work Areas

Now we turn to conditions inside the plant. If we are in a new facility and have the run of the place, we can declare the whole site a work area within our control. If we are in an existing plant, we make a designated work area. In either case, we establish uniform requirements upon entry to a work area and establish them with signs and/or barricades, examples:

Hardhats, safety glasses, hearing protection and closed toe shoes required.
Entry prohibited, escort required.
Entry prohibited, safety training required
Beware of welding flash
Tobacco Prohibited
Beware of Traffic
Beware of overhead work
Keep inside designated aisleways

We clearly define the work area so that the workers and visitors tune into a higher level of awareness as to their surroundings. Further, we do not expect to see any subcontractor personnel *outside* the designated work area when he is supposed to be inside working. Next, within the designated work area we establish specific "Laydowns." Let's imagine our hypothetical plant we laid out earlier on a 30' x 42' column pattern, each 1260 square foot area can be called a bay. We designate each contractor one, two, or three bays as space allows or as needed, for his or her exclusive use. Here, the contractor will store his materials, maintain an administrative area for paper and timekeeping, and a plan table. The contractor's laydown is his space for the duration of the job. We do require that it be kept neat, tight, and broom-swept at the end of the day. A proposed laydown plan for all contractors should be a point of discussion during the bid process. Perhaps the contractor will need a laydown at several locations, both inside and outside the plant. We go through this step to build cooperation prior to construction and avoid claims and consternation after the work has started. If the course of construction dictates that a laydown be moved, we, in accordance with of axiom 2 have planned for it and the contractor has included the cost of the move in his bid. We also have to consider whether the contractor's equipment can be parked within his own laydown overnight. Many lifts need to charge with 110V AC overnight and we need to make sure there are enough receptacles available. Contractors or Construction managers will need to provide outlets or pigtails to keep everything charged.

Receiving

Next, we designate a receiving laydown for inbound equipment and materials near the loading dock. The benefit and use of such a space depends on a number of factors. Remember how we made it easy to work with the procurement department by doing the harder, less routine part of the job? In our existing plant we encounter the same issues with the receiving department. Suppose we have two people; let's name them helpful, and not helpful. Suppose further they work in a bottling plant. On a daily basis they receive cardboard boxes for shipping, plastic bottles for filling, bottle-caps, pallets of sweeteners, other flavorings, etc. In all, 10 items on a routine basis. Helpful is new on the job and is excited to see the pant expanding because it makes for future job security. Not helpful has been on the same forklift for twenty years and the project is one more headache before retirement.

We have prepared uniform shipping instructions (as a general condition) for the project that go like this:

Ship to:	ACME Beverage
Street Address:	Arizona USA
Project:	Rocket Sled
Attn:	Contractor 1,2,3,4…or Owner or Vendor (a,b,c…)

And we have claimed an 84'x30' double bay which we tape off or paint the floor to like this:

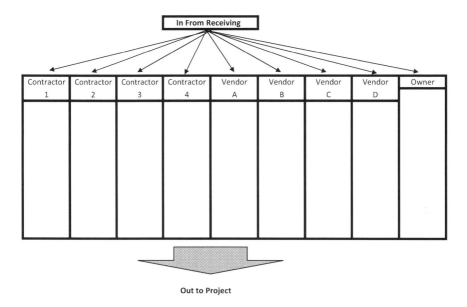

Out to Project

What we have done is create a project mailbox for most all of the non-routine inbound freight. Whether helpful or not helpful, the receiving department guy need not be concerned with the contents or final destination of the shipment. He merely looks at the label and moves it to the correct spot on the floor. The only other instruction needed is a call to the construction manager or owner's rep to let them know something has arrived. Vendors seem to have a knack to send things to the site before they have representation. And keep in mind, we have the riggers on hand to handle the big and heavy items. We give the receiving department the CM's cell phone number so he is the single POC for all in-bound freight, (and the first to know when something arrives) and so that not-so-helpful only has to worry about one phone number, even though he has the phone list on his computer.

Maintenance of Roads

Now, we need to consider the maintenance and repair or roads. What is the condition of the concrete and asphalt at the start of the job? Heavy trucks turning on asphalt can cause lots of damage in a short period of time. We anticipate a budget for repairs after the job or build it into our project scope.

Another thing to consider is whether we need to build and maintain any temporary roadways. How is the subsurface? How much rain do we expect? What kind of aggregates are available? Do we need a provision for ongoing dust control? If we do not plan for our road conditions, we may very quickly wind up with all contractors complaining about site access. And while it is a shared, common, and general condition, it's just not their problem.

Sanitation and Potable Water, Ice, and Use of Plant Facilities

Here is another shared and general condition that require planning and management. There may be no existing toilets, or a limited number far afield. The Owner or CM should handle all portable sanitation under a single account. If left to the subs, we might wind up with 3 or 4 port-a-pot contractors, with different levels of service, and an excess of inefficient sucker truck traffic. Potable water and ice should be permissible and available within the designated work areas and laydowns. We do not expect a worker to go without a drink of water between designated breaks, so it is better to provide coolers in the work area with a trash receptacle right next to them. If the plant allows the use of refrigeration, vending machines, cafeteria, lockers, etc., there should be some consideration as to existing capacity, housekeeping, and clean-up. Imagine a case where a plant employing 100 workers accustomed to wearing hairnets and dust-masks is suddenly flooded with 200 construction workers with a variety of backgrounds. It might very well be a better idea to erect a curtainwall or barricade to keep the plant and construction effort isolated and separate, if possible.

Tobacco Use

While not exactly a general condition, tobacco use is a general problem, so a few remarks are in order. If smoking is permitted in designated areas only, and the subs policy is during lunch and designated breaks only, we probably have a workable overall policy. We can specify ash cans filled with sand and emptied on a daily basis to accommodate the additional load. If the policy is no smoking on plant property, we can still expect a few miscreants smoking in their cars during breaks and we decide to look the other way or get rid of them. A headache all around, because skilled labor can be hard to replace. My basic opinion is that is better to give in and provide some additional designated areas, with metal pails filled with damp sand that can be changed out on a daily basis. Then, we enact a policy of zero tolerance outside the designated areas and break times.

Housekeeping

Good Housekeeping is the gateway to many positive effects: morale among the workforce, added productivity and a much lower probability for slip, trip, and fall accidents. Removal of trash and rubbish, crating materials, dunnage, and other wastes should be placed in suitable receptacles and /or removed from the site upon generation. Floors should be kept broom swept at all times and a second pass made with a sweeping compound. For larger sites, we should employ the use of a walk behind or drivable wet vac. I prefer to pay for a dedicated housekeeping force on a settled & separate budget under the general conditions. This force (minimum 2 or 3 guys, in support of 50 workers) support all housekeeping functions across all subcontractors. Waste is constantly moved from the work areas to the disposal area. We call these guys the Go-To team. When not moving the rubbish, they are sweeping up the slag, sucking up spills, cleaning the break areas or doing additional work in support of the plant housekeeping staff. One of the team has access to a work truck and can run for unexpected items and sundries. They can break open crates, operate a forklift, and take care of these shared, general conditions on the spot. A team like this will generally pay for itself by allowing the Journeymen to focus on the skilled aspects of their craft. We can usually get these guys from a temp agency for $25 or $30 an hour. Another alternate is if the job is done in the summer, we might have access to college students or children of plant employees who need a summer job. Let's pause and check our General Conditions budget.

(2) Go-To's, 40 hours per week, $25 per hour for 19 weeks = $38,000

Now take our (12) Journeymen @ $61 per hour, and give them the go-to budget:

$38,000 budget/$61/hr = 623 hours for housekeeping over 19 weeks, or 32 hrs/week or 3/12 =

2.7 hours/ week / man. More or less, we make everybody clean up for a half hour every day and have a clean site at the beginning of each day, but conditions deteriorate throughout the day as opposed to the continuously clean former case, and we lose 623 hours of productive journeyman effort over the same period. The former beats the latter.

Temporary Power

We discussed temporary power outside the plant for the trailers, but there may also be some requirements for temporary power inside the plant. Part of our planning process is to make that determination and prepare a general conditions budget if necessary. Our electrical contractor may have a panel board with a transformer and receptacles for use across the job that we might rent on a monthly basis. Such a board might plug into an existing 480VAC welding receptacle or spare breaker. Another thing to consider is whether we will have an outage for a power tie-in or transformer upgrade. If we need to rent a generator to keep the job moving, we plan and budget for the requirement.

The Weather and Snow Removal

One uncontrollable general condition is the weather. There is adverse weather such as rain and resulting soil conditions that can make it impossible to work outside. There are tables used by the Army Corps of Engineers based on data compiled by NOAA that are used to predict weather delays which are incorporated into their contracts; another way of setting mutual expectations to the best of the parties ability. Delays beyond the agreed to prediction should result in a day by day schedule extension for outside work. For inside work, the owner and the subs might agree that a snow day declared by the school district would provide grounds for schedule relief. Another stiffer trigger might be the declaration of a state of emergency closing the roads. The general condition of the snow on the site, once again, is an item for planning and budgeting. If the schools and roads are open, but the subs can't access the parking lot, they have a legitimate claim for relief. Snow removal on the site should remain the responsibility of the owner or construction manager.

Disposal

The final general condition we present for brief discussion is disposal. Construction will generate multiple streams of waste that need to be managed. Scrap is usually a subject of contention post-bid. Generally speaking, if the contractor is working on T&M or cost and fee, the owner has claim and title to the scrap, because he has paid for it. If the contractor is working on a lump-sum, he bid to install only the wire and aluminum required for a complete job and what remains belongs to him. That's the basic rule. Scrap ferrous metals will be generated by multiple contractors and may not be worth the cost of transportation or sorting out. Flip a coin to get rid of it, or if there is a profit, specify an end of job barbeque. The Owner or CM should provide sufficient roll-off service for the Construction and Demolition (C&D) waste. Let the Go-To guys pull the nails out of salvageable planks and lumber and set it aside, you never know when you might need some cribbing or a worker comes asking for some wood to build a chicken-house.

***

And just when you think we are capturing all of our costs, we must be clear that we have not. For years there has been a relatively high degree of confusion and inconsistency of method in capturing and specifying what to do with the cost of higher level supervision, bonds, and insurance. In other literature, these costs have been included within the general conditions. But as we have defined general conditions here and above, they clearly belong in a separate category we shall call "OVERHEAD." These costs will vary with the length of the schedule and the supervision requirements for the job. If the CM or owner sees the necessity for one or more of these positions to be full time and on the jobsite, he has the opportunity to specify the requirement and add a row or two to the schedule of values and call it project supervision. We present the following list to give the reader additional functions or duties that may be required by the scope and intensity of the work. Subject to further discussion and mutual agreement between the parties, we just declare that following are costs that may be incurred, but are not billable, unless otherwise specified.

Project Management:
Superintendent(s)
Safety Manager
CPM Scheduler
Quality Assurance / Quality Control
Project Manager(s)
Project Executive
Field Office Engineer
Field Office Support Staff
Project Expeditor

Assistant Superintendent(s)

Other Overhead:
Bonds and Insurance (excluding any for Subcontractors):
Builder's Risk Insurance
General Liability Insurance
Other General Project Insurance
Security, Payment & Performance Bonds
Bonds and Insurance Subtotal
ANY other home office costs beyond the oversight of the CM or Owner.

***

Finally, we allow the contractor to make his Profit. If the job is to be lump sum, we just add a line item for profit only, or, more typically roll up the project management, and other overhead above, into a category we shall call OH&P. If the job is on T&M or additions and deletions are done on T&M, the OH&P has been captured in the labor rates and the percentages we have agreed to for mark-ups on material and equipment. These details are presented in Chapter 6.

***

At this point in the development of our bid package, we are more or less complete for the piping portion of our job. We will now pay attention to some general requirements, submittals, forms and procedures that we incorporate in general, to this particular, and other projects down the road.

It will be presumed going forward that the usual procedure of sending out the package is followed by a pre-bid meeting and site walk-down. Questions are asked and answered, minutes to the meeting are taken and incorporated into the contract documents. In the grand scheme of things we will allow five weeks between the pre-bid and award, and two more weeks before mobilization, parsed out more or less as follows:

Week 1 - Send out the package
Week 2 – Pre-bid meeting and site-walk
Weeks 3, 4 – Contractors estimating period, back and forth with questions and RFI's

Week 5 – Bids due, evaluations, clarifications, final negotiations, Notice to Proceed

Week 6 – Execute contract, contractor acquires materials

Week 7 – Contractors mobilize and begin work.

If our reader did not realize the process would take this long, she has already committed a type one error. He is reminded that while our schedule of values used as an example was for the piping and insulation scope, we recall an electrical scope, some sheet metal work, and a scope for rigging and setting equipment. These are separate packages bid concurrently. For our purposes, we keep these jobs in mind, and we will continue to use the piping scope as the primary job for illustration going forward.

Chapter 5 Part 2. Receiving the Bid – Establishing the administrative and control documents.

The best laid schemes of Mice and Men
oft go awry,
And leave us nothing but grief and pain,
For promised joy!

-Burns

In our invitation to bid, we penciled in a period of performance for the work, setting forth our expectation *a priori* of how long it will take to get the job done. Recall this was part of the basis for our NPV analysis which justified the project in the first place. We have done most of our planning in accordance with axiom 2. We have done our best to keep the contractor away from a monopoly on his future power by building a true schedule of values. We turn again to this backbone document and place the tasks within the period of performance by requiring the contractor to submit a manpower loaded schedule with his proposal. The manpower loaded schedule submitted by the contractor is a REQUIRED continuation of the planning process.

Before we continue with the job at hand, let's introduce an exercise that will illustrate an element of time that most of us confront once a week or more due to lack of planning. We call it "missing the bus."

ASIDE

We must cross a bridge which is exactly one mile long in one minute. We can go as fast as we want, up to the speed of light, or 186,000 miles per second. We cross the first half of the bridge at 30 miles per hour.

How fast do we need to go to cross the bridge on schedule?

Close your eyes and think for a moment….. think. There is only one correct answer.

And that answer is "Not fast enough, because our time expired halfway"

Even when we could move at the speed of light.

END OF ASIDE

AXIOM 13

BEWARE OF GREEKS BEARING GIFTS AND THE CONTRACTOR THAT TELLS YOU HE CAN MAKE UP FOR LOST TIME BY ADDING A SECOND SHIFT.

Now, let's ask the same question from a different point of view, and examine what goes on in that portion of time that we have to begin with.

How fast do we have to go to cross a bridge one mile in one minute if there is a delay in getting started?

Let's use feet and seconds as our units. We have feet/second=5280/t, where t varies from 60 seconds at the beginning and ends with zero when our time is up. To keep the table short, we will look at our situation every five seconds, and ignore the representation of the end point (t=60) where we have already demonstrated that time is up.

We look at two derivatives in the division. The first one is the resulting velocity we must maintain the rest of the way to get the job done on time, and we observe that the required speed grows with each passing second. The second derivative, the change in the change, as each second passes is also increasing. Our problem grows at an *increasing* rate.

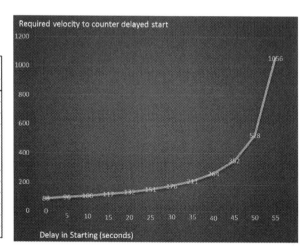

Delay in Starting Seconds	Required Velocity ft/sec	Second Derivative ft/sec
0	88	0
5	96	8
10	106	10
15	117	12
20	132	15
25	151	19
30	176	25
35	211	35
40	264	53
45	352	88
50	528	176
55	1056	528
59	5280	4224

Before moving to our manloaded schedule, we note that our speed across the bridge was one of those textbook examples that only exist in high school science class. We had no concept of losses due to heat or friction. Our velocity was only constrained by the speed of light. Further, we only considered a certain length of travel. Now, we must take this concept of velocity and add the human element we call productivity. How many bricks can a man lay in a day? How many feet of pipe can be installed, what is an optimum crew size – two working together are more productive than one working alone for twice the time; more often than not, three working together are more productive in 60 minutes than two working for 90.

Further, we add the concept of friction – too many working in a limited space slows everything down, unanticipated delays in materials, absenteeism, re-work etc. are working against the clock.

Here is a kind of shortlist of additional factors to consider as we move our hypothetical exercise of crossing the bridge to building the manloaded schedule from our actual schedule of values.

1) Every activity has a predecessor. It may or may not be true that a predecessor be 100% complete before successor activities may start.
2) The speed limit (man hours per day, either available for work, or actually worked) on a project can and will vary across the project.
3) In general work is not possible until material or equipment is available to facilitate the work.

4) The amount of work that can be done per man-hour of labor varies from situation to situation, e.g.: work done from lifts or scaffolds is less productive than work done on the floor.

5) A finite work space reaches a point of saturation, where the addition of workers diminishes overall productivity.

6) The work of one craft may not be available or desirable until the work of another is completed.

7) The nature of the construction will reside on a spectrum of both serial and parallel qualities. For example a high rise building can only start from one end. A road system can start at many nodes simultaneously.

8) A project may be bilinear. The English Channel tunnel and the Brooklyn Bridge were worked from both ends at the same time.

9) Creativity can have a positive impact on work at all times. When the transcontinental railroad was built through the Sierra Nevada, some of the tunnels were worked from both ends in, and, at the same time, from the center out towards both ends. Drilling a center shaft allowed the application of labor on four or more faces at the same time. The availability of work in the construction of the Interstate Highway system was constrained only by funding.

10) Technology can revolutionize the productivity of labor. The eventual use of nitro glycerin for blasting the aforementioned tunnels increased greatly the amount of rock that could be moved relative to prior use of black powder. In a personal example of not so long ago, I recall laying out a roadway with level-transit and a crew of three: one on the stick, one on the level and one on the tape and taking the notes. The job took six man-hours. Later the surveying engineer verified our lines and grades by taking a walk with a nine satellite GPS on a stick equipped with a data-logger. He worked without help and collected 10 times the data in all of 15 minutes.

11) As discussed previously, good housekeeping translates directly to better productivity.

12) Work which is available at one point in time might be done more or less productively than at another. For example, lights installed high in a ceiling may get done faster if there is no plant equipment directly in the way. Lower lighting closer to the ground might get in the way of the subsequent equipment installation, and more subject to damage…better to wait. Spend some time reviewing the drawings for interference. If we have cable tray and utilities piping installed in

the ceiling, done by two different engineers, check for interferences before the installation starts. Avoid rework. It is the CM's job to see it coming if the engineers missed it.

Let's return to a truncated version of our bid form:

			Process Utilities Scope of Work -Bid Form		Total Linear Feet	Labor Hours
Line Item	Area	Drawing	Scope	Reference Documents		
1	TA 1 and 2	M1.11	6", 4"and 1-1/4" from Pump Room wall near D9 to TA 1&2 plus 16 drops. Continue header to CL 4.		1297	1743
2	TA 1 and 2	M1.11	Insulation for Glycol Piping above		807	198
3	FCU 1	M1.12	Continuation of 1-1/4" from CL4 to FCU in MCC Room		125	115
4	FCU 1	M1.12	Insulation for Glycol Piping above		186	65
5	FCU 1	M1.12	Installation of Trane FCU and Ductwork in MCC room	FCU 1 15 2013 (MS Word), M1.15	NA	64
6	Pump Room	M1.13	Receive and set P1-P8, Chiller, Glycol Tank, Tower Water Tank, and HX1	LiDestri Foods Process Chillers (MS Word), Pump HX submittal (pdf)	NA	84
7	Pump Room	M1.13	All piping in Pump Room, and Between Room and Chiller. **Submit isometric for approval.**		427	2134
8	Pump Room	M1.13	Insulation for above		255	134
9	Roof	M1.14	Receive and set Cooling tower on existing support steel. **Submit transition piece drawing for approval,** fabricate and install transition. Complete tower piping not otherwise described above	Marley PF501130 P,M,S,G	60	110
10	TA 1 and 2	P2.11	Compressed air from tie-in at B6 to TA 1&2 and across rail shed	P5.01, P7.01, P0.01	720	449
11	Filling Area	P2.11	Compressed air from tie-in at B4 B-F, 2-3.5 (Drops as indicated)	P5.01, P7.01, P0.01	1080	340
12	Mix-Blend, CIP	P2.11	Remaining CA from tie-ins near B4, C5, and D5 (presumed) to Mix-Blend, CIP and drum stations	P5.01, P7.01, P0.01	710	479
13	Filling Area	P2.11	City water from tie-in at B2.5, to filling area (Drops as indicated)	P5.01, P7.01, P0.01	925	376
14	Mix-Blend, Almix, CIP	P2.11	Remaining CW from tie-ins at B3A, 4D(2) to Mix-Blend, Almix and CIP	P5.01, P7.01, P0.01	740	655
15	Mix Blend	P2.11	Install Nano water, complete as indicated.	P5.01, P7.01, P0.01	376	268
16	First Floor (all)	M2.11	Install all steam and condensate piping as shown	M5.01, M0.01, M7.01	1200	853
17	First Floor (all)	M2.11	Insulation for above	M5.01, M0.01, M7.01	2300	1120
18	First Floor (all)	M5.01	Installation of remaining steam / condensate stations, traps, etc.	M0.01, M2.11, M701	NA	
				Totals:	11208	9187

As we proceed, we keep our schedule of values side by side with the layout drawing. We can't put too much labor in one place at the same time. The craft required are:

Millwrights – Lines 6, & 9 to set the pumps (8), tanks (2) and heat exchanger (HX1) in the pump room and the cooling tower on the roof. These are predecessor activities to any associated piping

Mechanical Piping – We will let this be the steel piping for the utilities, lines 1, 3, 7, part of 9, as well as the steam and condensate piping lines 16 and 18. Note that the contractor returned NA for line 18. In subsequent discussions with this sub, she indicated that she recognized there was labor involved with line 18, and had incorporated those hours into line 16. A note was made of this phone conversation, initialed by the sub, and incorporated into the contract.

Plumbing Piping – This is the copper pipe used for the compressed air and city water Lines 11,12,13,14.

Insulation scope – Lines 2, 4, 8 & 17

There is a Nano water (reverse osmosis, perfectly aseptic water) system constructed of thin wall stainless, line 15.

And finally, an oddball FCU - (Fan Cooling Unit) lines 3, 4, 5, which involves hanging the FCU in the ceiling, some ductwork, mechanical piping, and insulation. A few hours by different crafts. We tackle this one by rigging it in place early, then the ductwork can be laid out and fabricated. Piping and insulation can be thought of as targets of opportunity, rather than forcing the work into a hard critical path.

Now, looking at the scope from this point of view, we have five main tasks plus our oddball. Experienced journeymen fitters could probably do any or all of the piping tasks, but we will organize them by common skill. Skills that arise from the use of common materials.

We have a copper team, a stainless team, and a "Black Iron" team - common vernacular for the team that welds and threads our steel mechanical pipe. A fourth team will be the insulators, and the fifth our riggers. The trivial amount of ductwork can either be done by our fitters, or by some journeymen out of the sheetmetal hall. It may be difficult, and surely inefficient to get two guys for just one day. Keep in mind that at least some installed piping is a predecessor to insulation work. We need not complete all of the piping before we begin the insulation. And also observe the differences in the installation rates per hour. We have, at our fingertips estimated production rates, so let's be sure to use them where we can. We will assert that the piping supports and hangers are included within the linear rates for each type of pipe

Look at Lines 1-4. We have a relatively long, fairly straight run of 1-1/4", 4" and 6" pipe, all black iron:

1297+125 feet of pipe =1422 feet at expense of 1743 + 115 hours = 1858 hours, or 1422/1858 = .76 feet per hour.

Corresponding insulation for the pipe in this area comes to 807 + 186 = 993 feet in 198 + 65 = 263 hours or 3.77 feet per hour.

So the insulation in this area proceeds 3.77/.76 = 4.96, call it, - five times faster than the pipe. Wow! Now, our first true scheduling issue arises: how shall we make sure we can keep the insulators busy once they are mobilized.

The corresponding situation in the pump room is a little different. The area is congested, the pipe diameters are bigger, and not all of it requires insulation. Bigger pipe takes longer for both the piping and the insulation:

Line 7 427' in 2134 hrs = .20 feet / hr for piping.
255' in 134 hrs = 1.90 ft / hr for insulation.

We expect the remaining pipe can be installed at the following rates:

CA line 10	1.5 ft/hr
CA line 11	3.2 ft/hr
CA line 12	1.5 ft/hr
CW line 13	2.5 ft/hr
CW line 14	1.1 ft/hr
NANO line 15	1.4 ft/hr

Steam and Condensate line 16 1.4 ft/hr
Insulation for the steam, condensate and most of the City Water Line 17 2.0 ft/hr.

We begin the development of our manpower loaded schedule, henceforth (MPLS) using Microsoft excel. Other scheduling programs are widely available and have more bells and whistles than most people can understand or use to advantage, but I will stick with excel, because you already have it on your laptop and probably know how to use it. Recall we are only capturing work per unit time and putting into a box, which is confined more or less by time, space and budget.

For brevity and simplicity, we begin by building a template or frame in accordance with our schedule of values and team structure (vertical axis) and place it against time (horizontal axis). Our first pass comes out something like this:

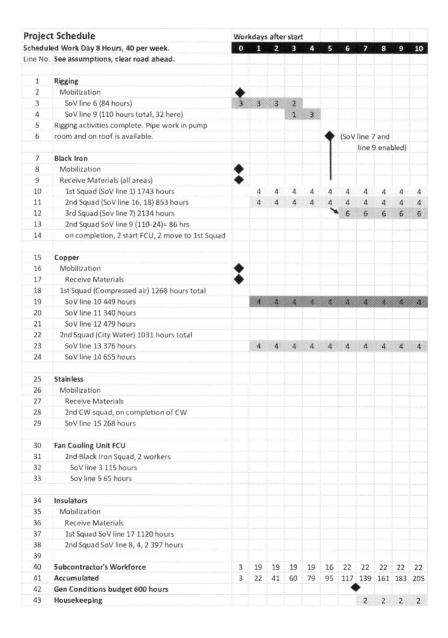

Project Schedule

Scheduled Work Day 8 Hours, 40 per week.
Line No. See assumptions, clear road ahead.

Line No.		Workdays after start										
		11	12	13	14	15	16	17	18	19	20	21
1	~~Rigging~~											
2	~~Mobilization~~											
3	~~SoV line 6 (84 hours)~~											
4	~~SoV line 9 (110 hours total, 32 here)~~											
5	~~Rigging activities complete. Pipe work in pump~~											
6	~~room and on roof is available.~~											
7	**Black Iron**											
8	~~Mobilization~~											
9	~~Receive Materials (all areas)~~											
10	1st Squad (SoV line 1) 1743 hours	4	4	4	4	4	4	4	4	4	4	4
11	2nd Squad (SoV line 16, 18) 853 hours	4	4	4	4	4	4	4	4	4	4	4
12	3rd Squad (Sov line 7) 2134 hours	6	6	6	6	6	6	6	6	6	6	6
13	2nd Squad SoV line 9 (110-24)= 86 hrs											
14	on completion, 2 start FCU, 2 move to 1st Squad											
15	**Copper**											
16	~~Mobilization~~											
17	~~Receive Materials~~											
18	1st Squad (Compressed air) 1268 hours total											
19	SoV line 10 449 hours	4	4	4	4							
20	SoV line 11 340 hours					4	4	4	4	4	4	4
21	SoV line 12 479 hours											
22	2nd Squad (City Water) 1031 hours total											
23	SoV line 13 376 hours	4	4									
24	SoV line 14 655 hours			4	4	4	4	4	4	4	4	4
25	**Stainless**											
26	Mobilization											
27	Receive Materials											
28	2nd CW squad, on completion of CW											
29	SoV line 15 268 hours											
30	**Fan Cooling Unit FCU**											
31	2nd Black Iron Squad, 2 workers											
32	SoV line 3 115 hours											
33	Sov line 5 65 hours											
34	**Insulators**											
35	Mobilization								◆			
36	Receive Materials								◆			
37	1st Squad SoV line 17 1120 hours								4	4	4	4
38	2nd Squad SoV line 8, 4, 2 397 hours											
39												
40	**Subcontractor's Workforce**	22	22	22	22	22	22	22	26	26	26	26
41	**Accumulated**	227	249	271	293	315	337	359	385	411	437	463
42	**Gen Conditions budget 600 hours**											
43	Housekeeping	2	2	2	2	2	2	2	2	2	2	2

132

Project Schedule

Scheduled Work Day 8 Hours, 40 per week.

Line No. See assumptions, clear road ahead.

		Workdays after start										
		22	23	24	25	26	27	28	29	30	31	32
1	~~Rigging~~											
2	~~Mobilization~~											
3	~~SoV line 6 (84 hours)~~											
4	~~SoV line 9 (110 hours total, 32 here)~~											
5	~~Rigging activities complete. Pipe work in pump~~											
6	~~room and on roof is available.~~											
7	~~Black Iron~~											
8	~~Mobilization~~											
9	~~Receive Materials (all areas)~~											
10	1st Squad (SoV line 1) 1743 hours	4	4	4	4	4	4	4	4	4	6	6
11	2nd Squad (SoV line 16, 18) 853 hours	4	4	4	4	4	4					
12	3rd Squad (Sov line 7) 2134 hours	6	6	6	6	6	6	6	6	6	6	6
13	2nd Squad SoV line 9 (110-24)= 86 hrs							4	4	4		
14	on completion, 2 start FCU, 2 move to 1st Squad											
15	Copper											
16	~~Mobilization~~											
17	~~Receive Materials~~											
18	1st Squad (Compressed air) 1268 hours total											
19	SoV line 10 449 hours											
20	SoV line 11 340 hours	4	4	4								
21	SoV line 12 479 hours				4	4	4	4	4	4	4	4
22	2nd Squad (City Water) 1031 hours total											
23	~~SoV line 13 376 hours~~											
24	SoV line 14 655 hours	4	4	4	4	4	4	4	4	4	4	4
25	Stainless											
26	Mobilization											
27	Receive Materials											
28	2nd CW squad, on completion of CW											
29	SoV line 15 268 hours											
30	Fan Cooling Unit FCU											
31	2nd Black Iron Squad, 2 workers											
32	SoV line 3 115 hours										2	2
33	Sov line 5 65 hours											
34	Insulators											
35	~~Mobilization~~											
36	~~Receive Materials~~											
37	1st Squad SoV line 17 1120 hours	4	4	4	4	4	4	4	4	4	4	4
38	2nd Squad SoV line 8, 4, 2 397 hours											
39												
40	Subcontractor's Workforce	26	26	26	26	26	26	26	26	26	26	26
41	Accumulated	489	515	541	567	593	619	645	671	697	723	749
42	Gen Conditions budget 600 hours											
43	Housekeeping	2	2	2	2	2	2	2	2	2	2	2

Project Schedule

		Workdays after start										
Scheduled Work Day 8 Hours, 40 per week.		**33**	**34**	**35**	**36**	**37**	**38**	**39**	**40**	**41**	**42**	**43**
Line No.	See assumptions, clear road ahead.											
1	Rigging											
2	Mobilization											
3	SoV line 6 (84 hours)											
4	SoV line 9 (110 hours total, 32 here)											
5	Rigging activities complete. Pipe work in pump											
6	room and on roof is available.											
7	**Black Iron**											
8	Mobilization											
9	Receive Materials (all areas)											
10	1st Squad (SoV line 1) 1743 hours	6	6	6	6	6	6	6	6	6	6	6
11	2nd Squad (SoV line 16, 18) 853 hours											
12	3rd Squad (Sov line 7) 2134 hours	6	6	6	6	6	6	6	6	6	6	6
13	2nd Squad SoV line 9 (110-24)= 86 hrs											
14	on completion, 2 start FCU, 2 move to 1st Squad											
15	**Copper**											
16	Mobilization											
17	Receive Materials											
18	1st Squad (Compressed air) 1268 hours total											
19	SoV line 10 449 hours											
20	SoV line 11 340 hours											
21	SoV line 12 479 hours	4	4	4	4	4	4	4	4			
22	2nd Squad (City Water) 1031 hours total											
23	SoV line 13 376 hours											
24	SoV line 14 655 hours											
25	**Stainless**											
26	Mobilization											
27	Receive Materials											
28	2nd CW squad, on completion of CW											
29	SoV line 15 268 hours	4	4	4	4	4	4	4	4	4		
30	**Fan Cooling Unit FCU**											
31	2nd Black Iron Squad, 2 workers											
32	SoV line 3 115 hours	2	2	2	2	2						
33	Sov line 5 65 hours						2	2	2	2	2	
34	**Insulators**											
35	Mobilization											
36	Receive Materials											
37	1st Squad SoV line 17 1120 hours	4	4	4	4	4	4	4	4	4	4	4
38	2nd Squad SoV line 8, 4, 2 397 hours								4	4	4	4
39												
40	**Subcontractor's Workforce**	26	26	26	26	26	26	26	30	26	22	20
41	Accumulated	775	801	827	853	879	905	931	961	987	1009	1029
42	Gen Conditions budget 600 hours											
43	Housekeeping	2	2	2	2	2	2	2	2	2	2	2

Project Schedule

Line No.	See assumptions, clear road ahead.	44	45	46	47	48	49	50	51	52	53	54
1												
2	Rigging											
3	Mobilization											
4	SoV line 6 (84 hours)											
5	SoV line 9 (110 hours total, 32 here)											
6	Rigging activities complete. Pipe work											
7	in Pump room and roof is available											
8												
9	**Black Iron**											
10	Mobilization											
11	Receive Materials (all areas)											
12	1st Squad (SoV line 1) 1743 hours	6	6	6	4							
13	2nd Squad (SoV line 16, 18) 853 hours											
14	3rd Squad (Sov line 7) 2134 hours	6	6	6	6	6	6					
15	2nd Squad SoV line 9 (110-24)= 86 hrs											
16	on completion, 2 start FCU, 2 move to 1st Squad											
17												
18	Copper											
19	Mobilization											
20	Receive Materials											
21	1st Squad (Compressed air) 1268 hours total											
22	SoV line 10 449 hours											
23	SoV line 11 340 hours											
24	SoV line 12 479 hours											
25	2nd Squad (City Water) 1031 hours total											
26	SoV line 13 376 hours											
27	SoV line 14 655 hours											
28												
29	Stainless											
30	Mobilization											
31	Receive Materials											
32	2nd CW squad, on completion of CW											
33	SoV line 15 268 hours											
34												
35	Fan Cooling Unit FCU											
36	2nd Black Iron Squad, 2 workers											
37	SoV line 3 115 hours											
38	Sov line 5 65 hours											
39	**Insulators**											
40	Mobilization											
41	Receive Materials											
42	1st Squad SoV line 17 1120 hours	4	4	4	4	4	4	4	4	4		
43	2nd Squad SoV line 8, 4, 2 397 hours	4	4	4	4	4	4	4	4	4		
44												
45	**Subcontractor's Workforce**	20	20	20	18	14	14	8	8	8	0	0
46	Accumulated	1049	1069	1089	1107	1121	1135	1143	1151	1159	1159	1159
47	**Total Man-Days 1156**											
48	**Total Man-Hours 1156 x 8 = 9248**											
49												
50	**Housekeeping**											
51	**Gen Conditions budget 600 hours**											
52	Mobilization	2	2	2	2	2	2	2	2	2	2	2

Project Schedule

Scheduled Work Day 8 Hours, 40 per week.
Line No. See assumptions, clear road ahead.

Workdays after start
`55`

Line No.	Description		
1	~~Rigging~~		
2	~~Mobilization~~		
3	~~SoV line 6 (84 hours)~~		
4	~~SoV line 9 (110 hours total, 32 here)~~		
5	~~Rigging activities complete. Pipe work in pump~~		
6	~~room and on roof is available.~~		
7	~~Black Iron~~		
8	~~Mobilization~~		
9	~~Receive Materials (all areas)~~		
10	~~1st Squad (SoV line 1) 1743 hours~~		
11	~~2nd Squad (SoV line 16, 18) 853 hours~~		
12	~~3rd Squad (Sov line 7) 2134 hours~~		
13	~~2nd Squad SoV line 9 (110-24)= 86 hrs~~		
14	~~on completion, 2 start FCU, 2 move to 1st Squad~~		
15	~~Copper~~		
16	~~Mobilization~~		
17	~~Receive Materials~~		
18	~~1st Squad (Compressed air) 1268 hours total~~		
19	~~SoV line 10 449 hours~~		
20	~~SoV line 11 340 hours~~		
21	~~SoV line 12 479 hours~~		
22	~~2nd Squad (City Water) 1031 hours total~~		
23	~~SoV line 13 376 hours~~		
24	~~SoV line 14 655 hours~~		
25	~~Stainless~~		
26	~~Mobilization~~		
27	~~Receive Materials~~		
28	~~2nd CW squad, on completion of CW~~		
29	~~SoV line 15 268 hours~~		
30	~~Fan Cooling Unit FCU~~		
31	~~2nd Black Iron Squad, 2 workers~~		
32	~~SoV line 3 115 hours~~		
33	~~Sov line 5 65 hours~~		
34	~~Insulators~~		
35	~~Mobilization~~		
36	~~Receive Materials~~		
37	~~1st Squad SoV line 17 1120 hours~~		
38	~~2nd Squad SoV line 8, 4, 2 397 hours~~		
39			
40	Subcontractor's Workforce	0	
41	Accumulated	1159	
42	Gen Conditions budget 600 hours		
43	Housekeeping	`2`	

Now, a list of assumptions and some overarching remarks are in order before we proceed to a general discussion on our man-power loaded schedule (MPL) Forgive me for not referring this as our People Power Loaded Schedule for the time being.

List of Assumptions:

1. We have 55 working days to get this job done, could be more or less within our constraints.
2. All administrative predecessors have been completed: contracts, permits, down payments. The road is clear.
3. Contractor's materials sufficient to support the start of efficient work are received on Day 1.
4. Capital equipment and Utilities Equipment discussed on prior chapters has already been installed.
5. Riggers mobilized prior to day 1, working to equipment installation, (2) above.
6. Installation of pipe supports are included in the time allotted for pipe installation.

Remarks:

1. The black diamond represents a "milestone"- a major culmination of predecessors leading up to the start or completion of substantial work. A milestone is a point in time and has no duration. Think groundbreaking or ribbon cutting as the most major milestones in any project.
2. Substantial completion under this schedule occurs on Work Day 49.
3. Given (2), We schedule the insulation to work *backwards* from a scheduled completion date on Day 52. Insulation can be worked starting anywhere along a completed line of pipe. We could possibly add many more workers closer to the finish date.
4. We pull 8 insulators out of a hat and put 4 on Steam/Condensate and 4 on the other systems. The line by line schedule for the insulation is arbitrary, and more or less interchangeable, worked for convenience.
5. Completing insulation on Day 52 allows three days total slack, to work unidentified minor items or punchlist.
6. The total man-hours on our MPL line 47 (9248) is a little higher than on our schedule of Values (9187). We have consistently rounded up to the nearest man-day when preparing the MPL, assuming a union show-up rule, or some other miscellaneous rework.

Now, let's look a little closer at what we have done. Each activity has a start date and a duration with a specific base of resources. We have taken our theoretical example of crossing the bridge and have pasted it into a very real and required scope of work. The scope has an exact length, e.g.: feet of pipe and a deadline for completion. We have set a speed limit (or pace) in which we expect the task will be completed. We know that we cannot travel at the speed of light, but for now we proceed within a 40-hour week in accordance with the expectations of our customer and workforce. Additional chapters on the nature of labor and the cost and benefits of overtime shall follow our present discussion.

***

When our project was laid before us in Chapter 1, the scope seemed enormous. We envisioned so many instruments, control signals, lists, and lists of lists. Our board of directors gave us the *"where"* and *"why"* of the project in the form of a power-point presentation before retiring to the golf course and leaving the rest to us.

In subsequent chapters, we got help. We added some engineers, we began to understand the problem of work and time, and the divisions of labor. We *discovered* resources like that guy we call Bob. We have begun to understand human nature and the economic behavior that we are likely to encounter with actors in the receiving and purchasing department. We have taken this amorphous scope of work and have specified the segments of work, the battery limits between them, the *"what"* shall be done. We have specified the materials, methods, and rules or codes to follow, the - *"how"* to do it. At last, we select our subcontractors, those *"who"* have responded to our request for a statement of qualifications, expressed an interest in our project, and have completed our schedule of values. Our subs have taken all of the foregoing into consideration and have reduced our effort into these elemental quanta we call a man-day; gathered them all up again, sorted them out and slapped them down on a calendar or clock we call the man power loaded schedule, our plan for *"when"*.

And now, having obeyed our second axiom, the planning is all but done. The grand alliance of cooperative subs and vendors is complete. We have committed to the schedule and budget and the owners have thrown their wallet on the table; still, we ask:

What could possibly go wrong?

CHAPTER 6

What specifically is a man-day, or man-hour, and what do we expect to get done with one.

Having constructed the man-loaded schedule from our estimate and schedule of values, we have made a pretty strong assertion as to how much labor will be required to get a more or less specific job done. We now turn to developing an understanding of how much the effort should cost in terms of dollars. Up to this point, we have been careful in the development of our bid package. We have made an exact specification as to the materials, methods, techniques, and quality requirements we expect to be used on the job, we have described the general conditions on the site and how those costs should be allocated, and we gave our subs an adequate amount of time to bid the job. As all these items were laid out before the bidders, and all having had equal access to material suppliers, the material, equipment, and general conditions, the cost of the job should be more or less equal among our competitors, except perhaps for labor.

We now arrive at a discussion of the base cost of labor, benefits, overhead and profit. Further, we set aside as "not labor" the fruits of people who might work at the controls of some capital equipment such as a long-wall miner that that might dig up 1,000 tons of coal in an 8 hour shift, or a continuous caster which might crank out a very expensive specialty steel like Inconel or Hastelloy at the rate of several tons per hour. We return to the labor classifications, which we discussed above, and for convenience we select a real sample of rates cut from an actual wage determination for part of Western Pennsylvania in 2015. These are typical for a 4th year Journeyman. Also known as prevailing wages, or Davis-Bacon (act of 1931 as amended) wages. These are the wages that *must* be paid on nearly all county, state, and federal projects, and also form the basis of labor cost for larger private construction utilizing union labor. As discussed earlier, a non-union worker might be paid less on a private project, but he or she would have different expectations as to the ongoing availability of local work as well as fringe benefits.

General Decision Number: PA150001 08/07/2015 PA1
County: Allegheny County in Pennsylvania.

	WAGES	FRINGE BENEFITS
ELECTRICIAN......................	$ 37.76	$ 23.17
MILLWRIGHT.......................	$ 37.35	$ 17.00
BRICKLAYER.......................	$ 28.95	$ 17.84
Asbestos Workers/Insulator....	$ 37.61	$ 22.82
BOILERMAKER...................	$ 42.67	$ 24.55

Hourly wages on the left are determined county by county, or some other political area, such as a group of counties, and arise from negotiation between the local labor unions and the contractors association.

The cost of "Fringe" benefits represent the cost of additional negotiated benefits that accrue *per-hour.* For example:

- Life insurance
- Health insurance
- Pension
- Vacation
- Holidays
- Sick leave
- Other "bona fide" fringe benefits

Just as important are some of the additional costs that *must* be paid, but are not counted as fringes. Some of these are payments required by federal, state or local law and are not fringe benefit contributions. Such payments required to fund Social Security, unemployment compensation and workers' compensation programs, as required by law, do not count as fringe benefits.

Is it starting to seem complicated? What could we possibly have left out?

One last remaining item for the job at hand is a provision for "Small tools and Consumables." We might have included a line item for them in our schedule of values, and just agree to a lump sum for everything needed to get the job done, or we might have agreed to pay consumables at cost and take title to the small tools remaining at the end of the job. Any of these schemes might work under the right circumstances, but there can be side issues with theft, agreement as to the quality of the tools required, and whether the owner or construction manager really wants to police the whole category and check that what is bought makes it as far as the contractor's gangbox.

To keep it simple, in our case we ask the contractor to set aside a percentage based on a labor hour to cover the cost of all small tools, and consumables necessary by craft type to perform the job efficiently. We can develop an extensive list if we want, and the contractor can strike a line through any item he feels offensive or not covered. The reader is reminded that we have already laid out a schedule of values for our project, and for the purposes here, we are constructing a further detail of what we will use to protect us from overcharges in the event of unexpected changes to our scope of work. We write of small tools and consumables only, not materials or hardware installed as part of the permanent installation, or equipment, used to improve the effectiveness of labor.

Think of it like this:

Materials: Anything installed on the job and left behind on completion including welding rods and hardware, or any scrap left on site or abandoned in place for convenience.

Equipment: Anything that is self-propelled such as lifts, trucks, or earthmoving equipment; or other power equipment costing more than $1000 such as welding machines and air compressors; or any non-powered equipment or instrumentation – optical levels, GPS systems, handheld multi-meters, lasers, etc. which are deployed on the project and taken away by the contractor at the end of the job, also including such things as scaffolding, & concrete forms.

Small Tools: ANYTHING that can be used by one or two people working by hand or costing less than $1000. e.g.- hammers, saws, grinders, burning outfits, screwdrivers, wrenches, drills, come-a-longs, chain and chain falls, rigging equipment –slings, shackles, all-types, capacities and lengths, jacks of all capacities and types chain vises, thread cutters, - wire rope in any configuration, pulleys of all types…. We could go on and on with this but the contractor should get the message in the first sentence above. Read it again.

Consumables: ANYTHING which is otherwise used up, expended, or discarded in the course of the work. All drill bits, saw blades, grinding wheels, rags, etc. All PPE: Gloves, safety glasses, hardhats, hearing protection, first-aid supplies, markers, pencils. Any and all types of FOGMA (Fuel, Oil, Gas & Maintenance) for any of the above tools or equipment. Any and all office supplies that might be delivered to the site for convenience.

And now, having further set our mutual expectations, we ask the contractor to present a cost on a per-hour basis for the small tools and consumables, and he or she returns something like $4-$8 per hour, depending on the job, and the labor classification. We complete the subject by taking note of the idea that a pipefitter might consume much more in the above category- oil, acetylene, oxygen, grinding and cutting supplies, than say a bricklayer that might go through a dust-mask a day and use the same trowel for a few years, or the operating engineer who sits in her air-conditioned cab all day and doesn't consume much of anything.

And we are just about ready to build our table of labor rates. We build a hypothetical sheet for an electrician, and recognize that we will have one sheet per class of labor. There are many templates like this in cyberspace. The key is to accurately capture each and every element of labor cost.

OWNER / CM	Glass Management, Inc.		
PROJECT NAME	Sony Expansion		PROJECT NO. 2015 - 22
CONTRACTOR	AA Electric		CONTRACT NO. 22-1
SUBCONTRACTOR			DATE 8/12/2015

HOURLY LABOR RATE WORKSHEET

(Reference 'Change Orders' in Contract General Conditions. Certified payrolls required for all workers on Project.
Contractor shall enter data into all fields highlighted in yellow; for fields highlighted in blue, data will automatically

	TRADE:	Electrician		CLASSIFICATION:			Journeyman

Item			Rate Per $100	Prevailing Wage Rate			Notes
				Regular Time	Overtime	Double Time	
Base Labor Rate				$ 37.76	$ 56.64	$ 75.52	Use certified payroll to verify
	Benefit Paid	Benefit Provided					
Fringe Benefits:	(put X in appropriate box)						
Pension [1]		X		8.65	8.65	8.65	
Health/Life Insurance	X			13.41	13.41	13.41	
Sick	X			0.70	0.70	0.70	
Vacation/Holiday [1]		X		-	-	-	
Other		X		0.41	0.41	0.41	
Fringe Benefits Subtotal				$ 23.17	$ 23.17	$ 23.17	
Total PW Hourly Rate				$ 60.93	$ 79.81	$ 98.69	= Base Labor Rate + Benefits Paid + Benefits Provided
Benefits Paid				$ 14.11	$ 14.11	$ 14.11	
Total Paid Hourly Rate				$ 51.87	$ 70.75	$ 89.63	= Base Labor Rate + Benefits Paid
Burden: Taxes & Insurance [2]							
FICA			0.0620	3.22	4.39	5.56	
Medicare			0.0145	0.75	1.03	1.30	
Federal Unemployment			0.0080	0.41	0.57	0.72	
Pennsylvania Unemployment			0.0400	2.07	2.83	3.59	Maximum - 0.062
Workers Compensation [1]			0.05	2.59	2.59	2.59	Usually less than 11%; can request policy
Other [1]				-	-	-	
Other [1]				-	-	-	
Burden Subtotal				$ 9.05	$ 11.40	$ 13.75	
Contractor Liability Insurance				N/A	N/A	N/A	Included in OH&P per CGC
Small Tools & Consumables				$ 7.00	$ 7.00	$ 7.00	
Other (warranty, record drawings, payment bonds, performance bonds, etc.)				N/A	N/A	N/A	Included in OH&P per CGC
TOTAL HOURLY RATE (Total Hourly Rate + Burden)				$ 76.98	$ 91.21	$ 112.44	= Total Hourly Billing Rate

Note: For change order work, mark-ups for overhead and profit shall be applied to the above rates (these rates are subject to audit) in accordance with the provisions of CGCs, under 'Change Orders'. Mark-up rates for utility repair work shall be adjusted in accordance with the CGCs, under 'Contractor's Responsibility for the Work', subsection 'e-Utilities'.

[1] Costs for Overtime and Double Time are same as for Regular Time.

[2] Taxes & Insurance apply to Total Paid Hourly Rate which includes Base Labor Rate plus benefits paid in cash.

By signing below, the submitter certifies and declares under penalty of perjury under the laws of the State of Pennsylvania that the foregoing is true and correct.

Rates certified by:		Company Name:	
	(print name)		
Signature:			

At this point, the bored or inexperienced reader will ask, why did we go through all this trouble to be so specific? The answer lies in our own protection. If the above exercise was left up to the contractor's designs, he would return a document unrelated to the workers wage and the contractor's cost structure. The worker is the only one that should benefit should a job be put on overtime – he or she is subject to less rest and leisure than customarily settled on the 40 hour week. Other costs and burdens are unaffected by the fact that we work more hours in a week and are paid at the same cost per *labor-hour*, not *labor-dollar*.

AXIOM 14

A GANG OF JOURNEYMEN GETTING OVERTIME DOES NOT ENTITLE THE SUBCONTRACTOR TO ADDITIONAL PROFIT, NOR DOES IT TAKE MORE OVERHEAD TO PROCESS A 10 HOUR TIMESHEET THAN ONE WITH 8 HOURS ON THE SAME PAPER.

On the bottom line, we arrive at acceptable billing rates - $76.98 per hour covers the full cost of straight time in a 40 hour week for our electrician who is fully equipped with tools, and provided with all of the tools and kit required to work in conditions of continuing good health and safety. The $91.21 and $112.44 rates are also appropriate. Our time and a half multiplier of 91.21/76.98 =1.19 and our double time multiplier of 112.44/76.98 = 1.46 are within the reasonable range of expectation and remove the incentive on the part of the contractor to game his way into an overtime condition.

Some of the more nefarious contractors we may cross down the road may say, "look, this is a lump-sum job and my rate structure is my business", but the union and non-union contractors are fundamentally different. We already noted that there could be a large difference in the cost structure between the contractor that has a safe work history, and another that does not. This is our business. These guys and gals, their safety record and the subs attitude in regards to the latter is a line item that we previously called the experience modification rate or (EMR). This cost is captured in the workers compensation line on the form above and we shall go out of our way to conclude a satisfactory verification. The non-union contractor, using travelers, or non-union workers not subject to prior agreement, has every incentive to cheat some profits into the very first line. He or she will be warned that the base wages are subject to audit and we do not expect the common shenanigans whereby they claim a base rate of $37 while paying $25. *There is a line item for overhead and profit elsewhere.* Using the rate calculation above, we strive for honesty, transparency, and accountability, including the right to audit contractor's payroll. They are, after all working for us, notwithstanding on a temporary basis. When changes roll around, we have every right to know the nature of the additional costs. Another thing we have done by establishing this control document is to avoid the claim the more odious contractors will assert: if changes are made which increase our scope of work, they *must be done* on an overtime basis, after the eight hour day is up, or sometime beyond the 40 hours. The claim is usually made by the non-union contractor who may not have ready access to additional labor in the form of more workers, and just finds it convenient to make the same guys work longer. We built the man-loaded schedule to fully disclose when and where additional work might be fit in on a straight time basis. The sub may need a little reminder from time to time: first, we shall add workers whenever possible. There is much more on the subject of overtime in Chapter 11.

Chapter 6. Part 2. More on Labor and Productivity.

In our previous chapter, we introduced the concept of efficiency of labor. Recall the list of items I referred to as friction or constraints working against productivity. We also examined some of the linear footage rates assumed in our schedule of values that the subcontractor used in developing his manpower requirements for the job. There are also virtually limitless production and cost data available from RS Means and other sources. Still, the estimation of labor requirements and productivity remain an art, rather than a pure science. Estimators do the best they can with the best information available, and in general, come up with pretty good results when compared against a statistical mean and standard deviation for a measureable effort. Yet, we cannot do a meaningful Six-Sigma on human labor subject to forces of nature. Some tasks are more predictable than others. Indoor, climate controlled work is more predictable than outside work. Repetitive work, such as floor tiling, painting, bricklaying, hanging lights, etc. is less risky than unclassified earthwork. Learning curves are real and exist and can substantially affect cost and schedule. And finally, work which is constrained by multiple predecessors, or a complicated supply chain, carry a greater degree of risk in getting started in the first place. The Central Pacific Railroad, built from Sacramento eastward required materials delivered by sea around Cape Horn. Despite the fact that progress was limited by grading and tunneling, through the Sierra Nevada, the project still suffered from lack of nails and rails due to shipping losses from time to time. The Union Pacific, working westward from Council Bluffs Iowa, enjoyed materials delivered from sections of track already completed from suppliers along the way. On a roll, the Westward effort could set a rail every 30 seconds.

***

In the end, we remain with our man-power estimate, start date, and completion date: the former developed by the subcontractor, the latter mutually agreed to. The start date, at risk of completion of funding, permitting, and all other preconstruction activities, might already be delayed by forces beyond our control, and the finish date unmovable: set at the highest level of management by those guys in the golf carts way back in our first chapter. We have done our best to adhere to axiom 2, and have done a pretty good job so far in understanding our scope of work and the capabilities and limitations of our vendors, suppliers, and subcontractors. We have communicated our intentions to work with no nonsense, shenanigans, or other forms of nefarious behavior. We have communicated our requirements to the best of our ability and the subs have returned bids, the man loaded schedules and labor rate sheets that indicate the communication has been two-way, five by five. We have completed 80% of the 65% effort called planning. What, at first, seemed an insurmountable effort has been reduced to many lists: Equipment, Utilities, Instrumentation, Controls, Vendors, and Suppliers. We have constructed maps of where things are to go. We have the drawings and specifications that show exactly with what and how things are to be done. We have made things easier for the plant personnel by not asking too much out of their routine. While we only illustrated one man loaded schedule for the piping work, we received others, electrical, ducting, and rigging. We have studied them extensively and have taken steps to assure that the various subcontractors will work un-fettered by the work requirements of the others. We have had a few joint discussions over the MPLs with all of the subs and have obtained buy-in among them. We have begun to build a spirit of cooperation in the heavenly direction of *teamwork*. The mutual expectations among the customer and between each and every other actor on the project have now been set and we are almost ready to get to work. The Preconstruction activities are complete, and before we move to part 2, (On Construction), we will pause for a short discourse on the subject of document and communication controls. We need to establish a chain of command and channels of communication, as well as an organized system for document maintenance and control.

CHAPTER 7

Controlling the beast we're about to unleash. Setting up the construction office and controlling the documents.

We now turn our guns on the subject of document and communication controls. We will attempt to draw a line between what is and is not necessary to control, and ask the question *"do we really need this on paper?"* Creating too much paperwork can be time consuming, but on the other hand, misuse of verbal communications can be disastrous. For the time being, I will err on the more conservative side. The evolution of communications by electronic means has provided faster and cheaper communications, but not necessarily rich of accurate or useful information. We must also make a declaration or redefinition that *"a document is any form of communication that may be printed on paper, be it electronic or otherwise, notwithstanding electronic origin or writer's original intent."* We must control the intended use, content, and distribution of most documents. Documents help make the rules for action and provide a legal record for dispute resolution.

If you are old school, go ahead and make a paper file for everything. If you are new school, you can specify: *"all official documents shall be presented in the form of a pdf or other approved format."* It is up to you or management at the time of this writing. Use no paper and keep it all on a suitably backed up laptop if you want.

Still, a great deal of care must be taken to ensure accuracy and clarity in what gets written and distributed.

Let's review the documents we have accumulated to date and add a few more. We have accumulated enough paper and/or electronic files that we require some discussion on their control.

Drawings
Specifications

Contract
Vendor Catalog Cuts (Cut Sheets)
Minutes to Pre-bid meeting
Bids
Schedule of Values
Manloaded Schedule
Master Schedule
Equipment List
Equipment Delivery Schedule
Utilities List
Instrumentation List
POC Telephone List
I/O List

To some extent, these documents, with the exception of the drawings, man-loaded schedule and schedule of values, can be considered to have reached a degree of completeness, finality, and acceptance. Other "living" documents have grown old at this point and may be placed in the appropriate file. The only thing that might make us pull them back out would be an unforeseen error or omission; the probability of such occurrence being in direct correlation to the amount of attention we have paid to Axioms 1 & 2.

A few of these documents will be referred to and talked over several times a day, and we still elect to print and pin copies to the walls of the construction office. It remains much easier to talk over paper spread out on a table, as opposed to looking sideways at a laptop. The documents of broad interest we should usually print and pin on the wall of our office are these:

Site Plan
Plant Layout
Designated Laydown Areas
Process and Instrumentation Diagram
POC Telephone List
Master Schedule
Equipment Delivery Schedule
Map to the Nearest Hospital (On the wall, copy in every gangbox)

We also take the time to print a paper set of all the drawings in "D size" - 22" x 34" and place them on drawing sticks according to trade or specification section: Plumbing, Mechanical, Electrical etc. There might also be a civil / structural / architectural stick and/or sticks that may emerge for the different varieties of equipment we have purchased. Here we define a "stick" as a hanging file of closely related drawings. There are many others, such as the GPS on a stick mentioned earlier.

ASIDE ABOUT STICKS, (for drawings and other uses)

Years later, on another job, the same Wildcat millwright, first presented in chapter five who told me he could not install a solenoid valve because it had those pesky *"waars"* sticking out of it suffered the misfortune of one of his workers accidentally dropping a hardened bolt down the throat of a very expensive and precisely machined and chromium plated extruder/screw assembly which would have been destroyed on start- up unless the bolt was extracted. Facing a very expensive and time consuming operation to tear down the 5 ton assembly, he walked into the construction office and asked if I had a "magnet on a stick." A quick review of the contract list of small tools and consumables revealed he had no duty to supply such a tool. Being more specific, I replied, "no, but I do have a 2 pound maximum lift magnet on a 48" telescoping aluminum rod with double insulated plastic grip." After raising one hand to his to his chin, his eyes directed upward, and presenting a silent countenance of deep ponder for a moment he replied "that might do the trick." A few minutes later he returned with the bolt in hand and said "Thanks for lending me this. Good thing that darned bolt had some *"Arn"* in her, she stuck right to your stick."

More Specifically:

"Arn" – Also known as Iron, a natural element having atomic number 26, Symbol (Fe) from the Latin *Ferrum*. Also happens to be the most common element on the face of the earth. Not to be confused with *"waars"*, or wires, usually comprised of the element Copper, atomic number 29, less abundant, more expensive, symbol Cu, derived from the Latin Cuprum.

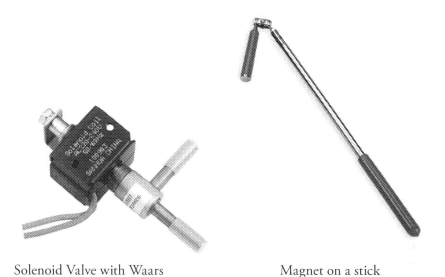

Solenoid Valve with Waars   Magnet on a stick

END OF ASIDE

PART 2 ON STICKS

Now, back to our drawings. Most experienced contractors and engineers have worked with the aluminum type of sticks with the wingnuts that clamp the drawings and fit on a custom designed rack. A system that might cost $500 to $1000. Another way to skin the cat is to use 1/8" x 1-1/4" lath strips of wood cut 6" longer than the width of the drawings to be hung (28" in the case of a D –size drawing), held together by 2 or 3 binder clips and placed on 2- "L" shaped shelving brackets affixed to the wall about 24" apart. This system is less expensive by an order of magnitude, and it is easier to add, remove, or replace a sheet for duplication by releasing the binder clips as opposed to loosening the wingnuts.

Having just saved the reader the cost of this tome with this simple yet elegant stick system, we continue on.

New documents are necessary and will emerge as we move to construction. A brief list follows:

Request for Information (RFI)
Revisions to Drawings
Field Orders
Change Orders
Construction Bulletins
Request for Progress Payments
Contract Letters
Vendor submittals
Daily Reports

Of course, in anticipation of these documents, we have already prepared appropriate sub-files in our project file; we further recognize that most of these documents will be specific to each subcontractor, so we create a subfolder for each one. RFI's during the pre-bid period will be communicated to each bidder, so we have only one folder for Pre-bid RFIs. During construction, an RFI is more likely to be subcontractor specific and there is usually no need to communicate the answer to everyone. Field orders are a document notifying the sub there is a change to his scope of work and are usually contractor specific. If a field order relocates a piece of equipment, several contractors may be impacted; they may have a positive, negative, or no change in their cost of doing the work. Contract letters, requests for payments etc. are normally considered to be confidential communications between the sub and the owner. Construction bulletins are site wide announcements that would be broadcast by e-mail or posted on a bulletin board. What I am driving at, is that the form of transmission, the distribution, the content, and the formality of the transfer of information are separate topics that merit some further consideration. Let's discuss a few of these documents and their appropriate uses.

How to ask and answer questions: The appropriate use and misuse of the RFI.

During the bid process, one of the first things we did was have a site walk and bid meeting. Representatives of each contractor had a chance to inspect the site and its general conditions. Verbal questions were asked and answered and discourse was recorded as a minute to the meeting. The meeting minutes were distributed the same day with an open window for comments, perhaps 48 hours. If no comments are received within that window, the document is attached *as is* as an exhibit to the bid and the contract. A comment may refine, refute, or request clarification. Further action on the part of the owner will be to acknowledge the comment with an appropriate response to all bidders and attach the comments to the minutes. We make clear that editorial comments are not desired and will not be incorporated into the minutes. At the close of the meeting, we make a statement to the bidders that verbal questions shall no longer be answered, and we set down some additional rules and procedures for subsequent questions during the bid period.

We expect that the bidders examine the entire bid package with the due diligence they would apply to any other significant matter of business. Make a statement in the bid meeting that "questions which indicate a lack of industry on the part of the bidder, or the answers to which are readily transparent or can be easily derived from the drawings are not desired. Repeated questions of similar nature by the same bidder may have a negative impact on final consideration of award."

Questions of a *general nature* such as what is the permissible load on the concrete floor or mezzanine deck, which are not called out in the documents may be asked by e-mail, and answered by the owner or construction manager with a simultaneous reply to all bidders; the e-mail string and/or similar e-mail strings being included as another exhibit to the contract. At this *general level* of the interrogative, we may permit more than one *directly related* question to be asked if the questions do not require referral beyond the owner, construction manager, or engineering discipline. More specific questions, which require further consideration, or indicate a real error or omission on the drawings call for the use of a more formal RFI.

An RFI is simply a formal *framework* or *protocol* for answering a single specific question. If the owner or construction manager cannot readily answer the question, we want to be able to route the RFI to a single engineer, architect, or equipment vendor.

We use the RFI when the question:

1. May be difficult to answer and take some time, possibly causing a delay.
2. May indicate a change to required materials, labor, and, subsequently, cost of the construction.
3. May result in a revision to the drawings and/or a field order.

4. May cascade to other scopes of work or require a "hold" on certain activities.
5. Should be recorded for the record.

It is the responsibility of the bidders to complete and submit the RFI on the day the question arises. Our aforementioned nefarious set of subcontractors may try to gang up a bunch of questions on different subjects and sandbag the owner by issuing the RFI at the last moment hoping for an extension of the bid date. The rare bidder of a more angelic nature might even suggest the answer to her question to help things move along: "drawing xxx-xx-xxxx indicates 4" pipe throughout drawing, yet the bill of materials calls out 6" elbows. 4" throughout seems correct, please confirm." The asked and answered period is also important to control. Of course, the owner and the bidder recognize that time is of the essence, in answering the question, but asking ASAP action on everything only serves to numb the process like the little boy crying wolf. Inasmuch as the owner and construction manager is responsible to have a full and complete understanding of the drawings, they may notice issues during the bid process and issue RFI's of their own. For example, the owner is responsible for his own floor loads and utilities requirements and, as mentioned earlier, must actively manage the battery limits and connections between the construction and equipment vendor scopes. If the construction drawing shows a 480 V 60 Hz 3 phase feed to a piece of equipment, and the equipment indicates a European type 400 V 50 Hz 3 phase load, that boundary MUST be managed by the owner or construction manager. A mechanical transition piece may not be that costly, but a big power transformer or frequency inverter might be relatively expensive and cause an overall delay to the schedule; a last minute type two error which shrouds in gloom all of the effort and good intentions previously embarked upon by everyone involved.

The rare and angelic contractor or vendor might have been able to point this out, but we probably did not issue the construction drawings to the equipment vendor or vice versa. The power problem slipped through the cracks because our foreign vendor deals with 400 volts all the time and would have thought a foreign concept to be supplied with any other. She may have thought the 60 Hz requirement was a typo on the utility list. Even if we wanted to attempt this degree of collaboration or sharing, the construction drawings were probably not available as a matter of timing when the equipment was purchased, nor could we reasonably expect the electrical contractor to be concerned with the equipment drawings (except for panel locations) beyond his general reference. These are kinds of *mental* battery limits; that cognitive space where one scope of *concern* intersects with another. Standard operating procedure for one party may be foreign concept to another. The RFI is a good tool to document ideas between the owner and the vendor that might have taken on some haziness along the way.

Having digressed far enough in our discussion as to the overall nature of the RFI, let's return to a specific model to solidify the concept. At a minimum, the RFI will get us to the Who-What-When-Where-Why of an issue with the same brevity of the first paragraph of a newspaper article.

Who is asking?
What specifically is the question?
When was it asked? When do we need to answer?
Where exactly on the drawings or in the specifications?
Why, because there appears to be an error, omission, discrepancy, or better way.

Here's a sample:

Request For Information ◯

(Number Assigned by Construction Manager)

ACME Chemical Company Butane Project

Question by:	ABC Electric
Specification Section Number:	16 Mechanical and 13 Electrical
Drawing Number:	ACME B 16-0-005 rev 0 and ACME B 13-0-005 rev 0
Drawing Reference:	Column Line 15, AB, Elevation 20'
Date:	09/10/15
Answer Requested by Date:	09/15/15

Question:

> The cable tray running East and West along column line AB collides with pipe rack running North and South along column line 15 at elev 20. Suggest cable tray offset to elev 22 to flyover pipe rack then return to elev 20. Please clarify.

> Recommendation is accepted, this RFI shall be incorporated into the contract. Bidders acknowledge and return with bid.

Answer:

Answered by:_____   Date:_____

Acknowledged by:_____   Date:_____

The RFI above seeks to clarify a simple *interference* in the construction. Interferences are worthy of the same attention given to battery limits throughout the course of design and construction. They arise from a variety of sources, may be hard to detect and manage early, and may or may not be cause for increased cost, depending on the timing of the discovery. Let's look at the example above in detail, and examine this collision of work in terms of the timing of the discovery, and place it somewhere in the spectrum of human behavior – nefarious on the left, innocence in the center, and angelic on the right. Picture a pipe rack carrying (1) - 8", (2) – 6" and (4) – 4" pipes running North and South 20 feet above the floor, and a cable tray running East and West at the same elevation, carrying 500 control wires, or a lesser number of heavy gauge power cables. The two are on a collision course just like two ships at sea or two vehicles approaching an intersection. Action must be taken or there will be a problem. We take a look at the timing of the discovery.

The best case is one in which our engineering contractor or construction manager saw it coming at the 75% review meeting prior to bidding and changed the cable tray or pipe elevation so the former would avoid the latter or vice versa. That's the way it should be, but recall our prior discussion where the mechanical guys were on one floor and the electrical gals on another, or separate buildings altogether.

A second case has been illustrated with the foregoing sample RFI. The discovery was made during the bid process and brought to our attention by the electrical contractor, who made a suggestion as to fixing the issue. He will include two off the shelf "S" transition pieces of tray and present a slightly higher cost for labor and materials in his bid, maybe $700

A third case might be one in which the tray has been completed, but the wires not yet installed. In this instance, a section of tray must be cut away, hangers will need to be modified, the transition pieces ordered on a rush basis, and the wire installation is delayed. We are led in the direction of a dispute whereby the electrical contractor asserts he must do the work on an overtime basis in order to keep the overall schedule….. Smells like $2500.

The worst case is when the tray and cables have been installed and the wires have been terminated at both ends. At this point, rework of the tray would be very costly. The wires would go to scrap because they are too short. Let's assume most of the pipe has been installed and welded together. Raising an entire section of pipe would be time consuming and require temporary suspension while hangers are re-worked, dropping the pipe to a lower elevation would require new hangers altogether, as the rods would be too short. We are faced with transitioning the pipe around the tray. Recall our rack carried seven runs of pipe: (1) - 8", (2) – 6" and (4) – 4" pipes for each, the bill of materials for the transition will require:

16 – 4" elbows
8 - 2' x 4" nipple
4 – 5' x 4" nipple

8 – 6" elbows
4 -2' x 6" nipple
2 – 5'x 6" nipple

4 – 8" elbows
2 -2'x 8" nipple
1 – 5' x 8" nipple

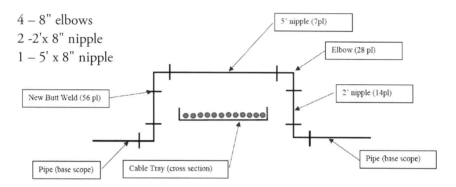

And just like that, the unfortunate and late discovery of this interference has introduced so much additional work that was unforeseen, unscheduled, and unbudgeted. Let's imagine two journeymen, all week to put this work into effect:

80 hours labor @ $84.00 (includes gas, rods, burning outfit)	$6720
8 hours designer @ $85.00	$ 680
Pipe Materials, Delivered	$1000
Subtotal	$8400
Overhead and Profit 15%	$1260
Total	$9660

Having touched from time to time on the sin of temptation, and other forms of rude behavior, we trust we have not qualified and invited to bid the type of contractor who might notice the interference prior to bidding and chose to remain silent. In fact he or she may have taken a red pencil and drawn a circle around the problem and made a note: "big change order later." These are your so-called "change contractors" – usually focused on poorly specified public works projects; they bid low to get the job then go after every change imaginable. This behavior can be expected in the public arena, where the bidding goes to the lowest bidder without exception, and there is no expectation of future work based on meritorious conduct. In our case, we have made every effort to exclude such a beast from our party by a thorough vetting we discussed in earlier chapters. Change-driven contractors seldom come with good reputation, satisfactory references, or a history of repeat business.

Other types of interferences are common and merit a brief discussion. Most typically, these occur (in the space domain) at the battery limit of the architectural work and the plant/process engineering, and/or (the time domain) where the schedule of the architectural construction and the equipment installation might collide. For instance, we might have on order a standard steel building or concrete tilt-up building with ventilation louvers in standard locations along the walls; subsequent equipment layout might overlook these louver locations and place equipment so as to restrict or otherwise alter the desired air flow, or a fan motor on the louver might be implied on a vendor cut sheet, but not be shown on the construction drawing. We have the same issues with the intended direction of door swings, etc. Typically overhead lighting on the architectural lighting plan might rendered useless having been blocked by subsequent installation of process equipment or a structural mezzanine deck. We shall take some extra care when locating lights and predicting shadows. Sometimes equipment may be too big to fit through a completed roll-up door opening, so we either hold up on completing the door, and move the equipment through a larger opening in an incomplete wall, or size the door to accept the equipment, keeping in mind it is easier to create a temporary opening in a steel building than in one made from pre-cast concrete.

Thoughts upon the issuance and control of drawings:

In the RFI above, we referenced a drawing. The number of which has a specific meaning, like a part number or catalog number. Let's take a moment to examine our meaning in this specific case:

ACME-B-16-0-005 rev0

Revision Number
Sequential Number in Area
Area Designation (Indexed across other drawings)
Construction Division 16 (Mechanical)
Project Designation (Butane)
Company Name

Our electrical drawing in the same area of the plant, at the same stage of development is ACME-B-13-0-0005 rev0. And we would have attempted to maintain the same numbering schemes across the other divisions of work.

At the start of the project, the project manager or project executive issues a block of numbers to each engineer/architect and they proceed to develop the drawings in accordance with a confined logical structure so as to avoid duplication and invite comparison between divisions. There are numerous schemes available, and I have just presented one of many possibilities. In this case, the very first conceptual drawing would be designated Rev A, and we would advance up the alphabet as revisions are developed, making notes of the existence of previous revisions in the title block. Each revision should be important enough to merit a different and progressively more exact meaning. Rev A – concept for approval, rev B preliminary, rev C - for approval, D, E, F…. depending upon the complexity of the effort, or lack thereof. On the issue of the bid set of drawings, all drawings are switched from alphabetical to numerical control. Rev 0 is always reserved for "issued for bid" or IFB. Post bid, it is good practice to make another issuance of the drawings, rev 1, "issued for construction," or IFC whether or not there has been a change between the rev 0 and rev 1 set. If we go back and reference our RFI which was asked during the bid period, we answer the question and issue a sketch like the one above showing the offset and our bill of materials. Sketch SK-B-16-005-1 is incorporated into the bid set of drawings, and our rev 1 drawing is revised and detailed to show the change, but we have captured the cost in the original bid.

As a final note of drawing control, let's note that we have issued the engineers a block of drawings for their use, and it is up to their team and managers to issue the revisions as things are developed. Beginning with rev 0; however, it must be the sole duty of the construction manager to issue subsequent drawings. We cannot allow our electrical engineers to issue drawings directly to the electrical contractor under uncontrolled circumstances. While we have specified that all files be transmitted as a pdf, it has become too easy to transmit drawings by e-mail and too easy to lose control of drawing content if the transmittal and intent goes uncontrolled. More and more, we see drawings that have the same exact title-block, but contain differences in content. If we lose control of what is an official and approved drawing, we have begun to lose control of the project. Starting with rev 1, all drawings must be issued by letter of transmittal, if there is no cost impact, or by a Field Order or FO if there is a cost or schedule impact.

On the one hand, there are projects where the same drawings are issued over and over again, using a standard design. A ranch house on one level acre, a Wal-Mart on 20, or a No.5 McDonalds with Playland option. For jobs like these, the same set of drawings might be used over and over again and might only get revised if an architectural finish is changed or discontinued. The drawings are issued to a GC and he does the rest on a lump sum, the work is clear and repetitive and there is no need for heavy involvement on the part of a construction manager or owner's rep.

On the other hand, our net present value analysis indicates very heavy initial returns, followed by rapid, decline and medium term (4-5 year) obsolescence. The faster the job gets done the more we will be rewarded. The engineering is 75% complete, but final design is lagging, our schedule indicates that bringing the production date forward will more than justify overtime. The press of business dictates that we move forward and anticipate changes/ additions. By anticipate, we mean we expect these things. Our lump-sum bid is returned at $1 million, and our budget for expected changes and additions is another $250,000 or $2.5 million on a $10 million base scope. We have a staff of engineers working to solve the remaining problems. We want them to work closely with the contractors, but we cannot allow out of control verbal direction in the field.

The field order is your new best friend.

The field order is our all-purpose control tool as the construction moves forward. This is a single sheet of paper that:

1. Directs and authorizes a change to the scope of work and lets the contractor know he can expect to get paid for it.
2. Authorizes overtime as needed, to new or existing work.
3. Instructs the contractor to (choose one):
   a. Proceed with the work on a time and material basis.
   b. Present a lump sum estimate before proceeding
   c. Acknowledge a zero-cost change or administrative only change to the scope of work
   d. Provide a credit for work that is deleted
4. Establishes an owner's order of magnitude estimate for the cost of the work.
5. Sets up a specific cost account for documentation purposes and or back-charges.

Let's have a look at a few examples.

In the first case, we have mobilized a contractor to install sanitary piping in an operating beverage plant. Remember our housekeeping go-to guys? The plant management wanted to use their in-house HR department to hire them, even though the routine there is to hire filling machine operators. Out of courtesy and political correctness, we said OK go ahead. The HR department, uncomfortable with our non-routine requirement, and never having heard of a go-to guy has instead hired a consultant to determine the appropriate wage scale and skill set for them. At best, they will join the team before the job gets done. It remains that we must remove a dust barrier / curtain wall that was left behind by another contractor who has since demobilized. One of our aforementioned changes that was captured in an Unidentified Minor Items (UMI) budget a long time ago. Because our go-to guys are not yet on the job, but out sanitary piping guys are, guess who gets to do the job? The last thing we need to do; however is verbally ask our sub to do us a favor. He just showed up as scheduled with another job to do. In fact, if you sent a young project engineer to make the request, he might just laugh at her. There is an old saying that nothing on a job takes less than 4 hours to do or costs less than $500.00

AXIOM 15

THERE IS NOTHING ON A JOB THAT TAKES LESS THAN 4 HOURS OR COSTS LESS THAN $500 IN 1995 DOLLARS, ADJUSTED FOR INFLATION, SO GET USED TO IT.

Still, we must remove our curtain wall before we do anything else, so we write one of our first field orders for the job:

Field Order A&B 002

LiDestri Foods Gemini Project

Issued to: A&B Process Piping
               December 13, 2013

Description of Work:

Remove curtain-walls at column lines 4 and 7 and install red tape around the new sloped floor. Complete the tape installation by close of business this afternoon and the curtain wall removal by 0700 Monday December 16. Fold the visqueen neatly and place it as directed by Dave Kulak. Return the red tape to the construction office.

Action:

☐   Do not proceed with the work. Provide lump sum price IAW the description.

☒   Proceed with the work in IAW the lump sum price submitted. Sign below when completed and return.

☐   Proceed with the work IAW the time and material provisions in your contract. Not to exceed engineer's estimate without additional notice. Return this Field Order with daily time sheets, material receipts and other documents required for payment

☒   This is for documentation of a zero cost change. Proceed with the work as noted. Sign below when acknowledged and return.

☐   Provide credit for the deleted scope of work in accordance with the change provisions in the contract.

☐   This field order extends end date of the contract to:

☐   This field order does not extend the period of performance of this contract.

A change order modifying the financial terms of the contract shall be issued on completion of the work as shown.

Engineers Estimate $500

DE Conver
Authorized by:

Completed by:_____Date:

And here's your first change document. Again, what, where, when, how to do exactly, and specifically what needs to get done. In this case, we don't need to document the why. The FO is executed as directed, our contractor now has a means to justify payment, and we have a written record of the work. Just like a service ticket you get at the garage when your car is worked on. In the background, maybe the plant maintenance manager has been trying to get the other contractor to come and remove this wall for weeks, but the job was done and they have been unresponsive. He now has a document to back-charge his final billing. Finally, instead of waiting for the other contractor to show up, and held hostage to his mercy in the meantime, we have effectively cleared our first, admittedly minor obstacle, and let our sub, and the plant know we are serious about getting the job done. We are taking the most important political step towards taking control of the project.

Here is a second example that illustrates good control and management over a battery limit. Two of our filling machines (coming from Sweden) came with cut sheets that showed a requirement for 20# per hour of steam delivered at 20 psi. Our plant steam is in a 4" header at 50 psi. Accordingly, our mechanical engineer designed a pressure reducing station between the header and our filling machines, and it was captured in a drawing detail that looked like this:

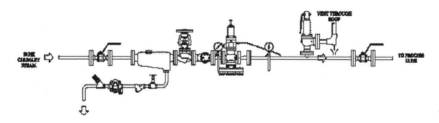

1 CULINARY STEAM PRESSURE REDUCTION DETAIL
NOT TO SCALE

Our mechanical engineer clearly specified the manufacturer of the valves, drip legs, etc. The contractor obtained his pricing from a local supplier and allowed four hours to make-up the flanges. Upon receipt of the equipment, we open the mechanical panel and discover a similar array of valves having been installed as part of the equipment package in Sweden. Somewhere along the way, the battery limit was misunderstood. A simple miscommunication that might have proved costly. Perhaps, if the cut sheet said something like: "minimum supply 20# at 20 psi, maximum inlet pressure 150#, this problem might have been avoided. In any event, and if we catch it in time, following the discovery of the duplication, we immediately issue a field order placing a hold on the purchase of any material associated with detail 1 above.

Now, let's discuss the timing of this discovery. Pre-bid, we just issue an addenda deleting the labor and material from the scope of work. All bidders take the same action and competition is maintained. If we discover the duplication post-bid, but before our contractor has ordered anything we can deduct the exact cost of the materials and the mark-up thereon, but will probably haggle or surrender on the cost of the labor to make it up. Finally, if we don't pay enough attention to even notice the duplication, we have just put the money we didn't know we had into the incinerator we talked about back in chapter 1.

In this case, our equipment engineer pointed out the duplication before the reducing stations were made up:

Field Order IWM3005

LiDestri Foods Gemini Project

Issued to: IWM3
              December 20, 2013

Description of Work:

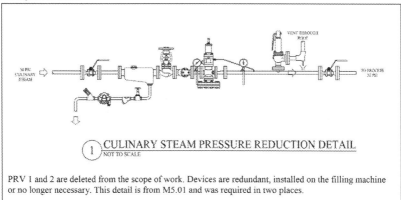

CULINARY STEAM PRESSURE REDUCTION DETAIL
NOT TO SCALE

PRV 1 and 2 are deleted from the scope of work. Devices are redundant, installed on the filling machine or no longer necessary. This detail is from M5.01 and was required in two places.

Action:

☒ Provide credit for the deleted scope of work in accordance with the change provisions in the contract.

☐ This field order does not extend the period of performance of this contract.

A change order modifying the financial terms of the contract shall be issued on completion of the work as shown.

Engineers Estimate: Materials cost less restocking charge if applicable. Labor IAW RS means.

Authorized by:

Completed by:_____Date:

In this case, the field order was returned with credit of $10,596.30

Our next example is a field order for work which was deleted following receipt of bids and procurement of materials, but prior to installation. We captured the scope in our schedule of values, line 15 which is reproduced here:

Process Utilities Scope of Work - Bid Form

Line Item	Area	Drawing	Scope	Reference Documents	Total Linear Feet	Labor Hours
1	TA 1 and 2	M1.11	6", 4" and 1-1/4" from Pump Room wall near D9 to TA 1&2 plus 16 drops. Continue header to CL 4.		1297	1743
2	TA 1 and 2	M1.11	Insulation for Glycol Piping above		807	198
3	FCU1	M1.12	Continuation of 1-1/4" from CL4 to FCU in MCC Room		125	115
4	FCU1	M1.12	Insulation for Glycol Piping above		186	65
5	FCU1	M1.12	Installation of Trane FCU and Ductwork in MCC room	FCU 1 15 2013 (MS Word), M1.15	NA	64
6	Pump Room	M1.13	Receive and set P1-P8, Chiller, Glycol Tank, Tower Water Tank, and HX1	LiDestri Foods Process Chillers (MS Word), Pump HX submittal (pdf)	NA	84
7	Pump Room	M1.13	All piping in Pump Room, and Between Room and Chiller. **Submit isometric for approval.**		427	2134
8	Pump Room	M1.13	Insulation for above		255	134
9	Roof	M1.14	Receive and set Cooling tower on existing support steel. **Submit transition piece drawing for approval,** fabricate and install transition. Complete tower piping not otherwise described above	Marley PF501130 P,M,S,G	60	110
10	TA 1 and 2	P2.11	Compressed air from tie-in at B6 to TA 1&2 and across rail shed	P5.01, P7.01, P0.01	720	449
11	Filling Area	P2.11	Compressed air from tie-in at B4 B-F, 2-3.5 (Drops as indicated)	P5.01, P7.01, P0.01	1080	340
12	Mix-Blend, CIP	P2.11	Remaining CA from tie-ins near B4, C5, and D5 (presumed) to Mix-Blend, CIP and drum stations	P5.01, P7.01, P0.01	710	479
13	Filling Area	P2.11	City water from tie-in at B2.5, to filling area (Drops as indicated)	P5.01, P7.01, P0.01	925	376
14	Mix-Blend, Almix, CIP	P2.11	Remaining CW from tie-ins at B3A, 4D(2) to Mix-Blend, Almix and CIP	P5.01, P7.01, P0.01	740	655
15	Mix Blend	P2.11	Install Nano water, complete as indicated.	P5.01, P7.01, P0.01	376	268
16	First Floor (all)	M2.11	Install all steam and condensate piping as shown	M5.01, M0.01, M7.01	1200	853
17	First Floor (all)	M2.11	Insulation for above	M5.01, M0.01, M7.01	2300	1120
18	First Floor (all)	M5.01	Installation of remaining steam / condensate stations, traps, etc.	M0.01, M2.11, M701	NA	
				Totals:	11208	9187

The field order follows:

Field Order IWM3023
LiDestri Foods Gemini Project

Issued to: IWM3 Mechanical
          February 13, 2014

Description of Work:

> This field order deletes the scope of work for the Nano water as shown on P2.11, and bid as item 15 on the schedule of values. R/O water pipe shown on M2.11 is also deleted by this order. Engineer's take-off is attached for reference.

Action:

☐   Do not proceed with the work. Provide lump sum price IAW the description.

☐   Proceed with the work in IAW the lump sum price submitted. Sign below when completed and return.

☐   Proceed with the work IAW the time and material provisions in your contract. Not to exceed engineer's estimate without additional notice. Return this Field Order with daily time sheets, material receipts and other documents required for payment

☐   This is for documentation of a zero cost change. Proceed with the work as noted. Sign below when acknowledged and return.

☒   Provide credit for the deleted scope of work in accordance with the change provisions in the contract.

☐   This field order extends end date of the contract for this work only.

☒   This field order does not extend the period of performance of this contract.

A change order modifying the financial terms of the contract shall be issued on completion of the work as shown.

Engineers Estimate: Credit $ (37,152.00)

signature
Authorized by:

Completed by:_____Date:

Now, the engineer's take-off revealed 524 feet of pipe, as compared to the contractor's estimate of 376 linear feet, so in this case, the contractor might have escaped a loss if the work went forward. Second, we allowed a 20% re-stocking fee on the material, the contractor having demonstrated his receipt thereof, confirmed by visual inspection.

As a final example as to the good use of a field order, we return to our thought that we need to manage the interaction between our contractor and the equipment engineers. In this case some last minute changes to some steam piping were taking place. There was a requirement for culinary grade steam going to a filling machine. The utilities mechanical engineer specified an expensive heat exchanger to heat the nano water, but subsequent investigation uncovered much less expensive culinary grade filters to serve the same purpose. While that change was occurring, we still needed to supply plant steam to the equipment while it was undergoing start-up and debug. Keeping in mind that we cannot allow uncontrolled communications between the engineers and contractors, we can direct the contractor to work at the direction of an engineer, within a limited budget and scope of work:

Field Order IWM3001
LiDestri Foods Gemini Project

Issued to: IWM3
            December 5, 2013

Description of Work:

> Provide labor and materials to run a temporary steam service to the filling line. Work from the permanent tie-in location on drawing M2.11, CL D, 5.5 at existing valve elevation. Run 1" as shown, to remain as part of the permanent contract work. Provide schematic sketch using number FOIWM3001 prior to any fit-up. Keep separate time, do not charge for 1" work in the contract. Terminate the supply on the filling line at the direction of Christer Forsberg. Work should proceed on straight time and conclude on Tuesday December 10 by end of shift. Overtime not authorized.

Action:

☐   Do not proceed with the work. Provide lump sum price IAW the description.

☐   Proceed with the work in IAW the lump sum price submitted. Sign below when completed and return.

☒   Proceed with the work IAW the time and material provisions in your contract. Not to exceed engineer's estimate without additional notice. Return this Field Order with daily time sheets, material receipts and other documents required for payment

☐   This is for documentation of a zero cost change. Proceed with the work as noted. Sign below when completed and return.

☐   Provide credit for the deleted scope of work in accordance with the change provisions in the contract.

☐   This field order extends end date of the contract to:

☒   This field order does not extend the period of performance of this contract.

A change order modifying the financial terms of the contract shall be issued on completion of the work as authorized.

Engineers Estimate $1,000 - $5,000

Authorized by:

[signature]

Completed by:_____Date:

Hopefully I have demonstrated the several and good uses of the field order. Keeping in mind that we have five or six contractors working in the same place at the same time, on a fast track schedule with incomplete engineering, we can have field orders popping up several times a day. The CM's main duty is to issue them as they arise, same day as the issue appears, and sometimes even before the contractor takes notice. This is the normal way we control our changes. We saw most of them coming and made a budget to the best of our ability. The key to the whole concept is that we avoid any and all horse trading, we have a discreet and compartmentalized document as to cost, a permanent record of the work. We thereby establish a measure of financial control over the contractor.

AXIOM 16

THE TAIL SHALL NOT WAG THE DOG

Finally, the contractor will appreciate the approach because it is up to him to complete the document for billing. Depending on the length and intensity of the job, the CM might issue a change order (CO) every other week. As field orders are returned, complete with timesheets, materials costs and any field sketches, we simply summarize the cost of each one onto a one or two page document (the Change Order) which changes the financial terms of the contract. In the foregoing examples, we have pulled documents from an actual job. Interstate Welding and Mechanical (IWM3) was our utilities mechanical contractor. Working this job over a period of about 90 days generated 33 field orders and a net reduction in the scope and cost of the work. A substantial amount of condensate pipe and nano water pipe were deleted. The actual change order follows. It is simply the sum of the changes issued as field orders.

Gemini Project
Contract Change Order Number 1
March 1, 2014

IWM3 Mechanical
8407 River Road
Pennsaauken NJ 08110

AMOUNTS BELOW DO NOT INCLUDE ANY TAXES WHICH MAY BE PAYABLE

Original Contract Amount: $ 656,000.00
Field Changes Completed to Date:

1	Temporary Steam for Filling	$	904.00
2	Admin Only - Pump Room Coordination	$	-
3	Provide Cardboard - Support General Conditions	$	500.00
4	Install supports for roof mounted equipment	$	2,767.40
5	Delete PRV Stations	$	(10,596.30)
6	Delete Condensate Pipe	$	(10,000.00)
7	Extend Compressed air over Packaging Line	$	6,278.90
8	Delete CA, Bid Item 10	$	(7,924.00)
9	Install air per sketch, replaces pipe not used in FO8	$	546.00
10	CA to mix blend and almix, delete stand	$	(1,500.00)
11	Not Used	$	-
12	Delete remaining condensate west of CLD	$	(40,814.00)
13	Relocate CW interfering with partition	$	1,365.00
14	Modify Flex connectors, P7 & P8	$	435.00
15	Changes to Almix piping	$	-
16	Delete Culinary Steam Generator, and add Filters	$	(12,512.00)
17	Transmittal of Torque Spec.	$	-
18	Not Used	$	-
19	Support for EDS (Install Bypass Valves)	$	803.75
20	Air and Water to second cardboard packer	$	2,260.19
21	1/2" Air to receiving	$	1,841.20
22	Add isolation valves to TA machines	$	4,359.72
23	Delete Nano Water from Scope	$	(23,331.00)
24	Add new nano scope	$	2,500.00
25	Complete 1/2" tie-ins at FCU	$	696.00
26	Booster pump (Not Used)	$	-
27	Gauges for Pumps in Mechanical Room	$	3,285.00
28	Vents for Chiller	$	4,803.72
29	Start-up support Dave Wallace	$	3,641.50
30	Nano Room (Not Used)	$	-
31	Chemical Drains for Filler	$	4,766.00

32	Isolation for CLA VAL	$	1,200.00
33	Relocate CW pipe to Fillers on wall to allow TA service	$	1,074.20
	Net changes to date (estimated)	$	(62,649.72)
Revised Contract Amount		$	593,350.28

This Change Order accounts for all changes to the contract amount on Substantial Completion. Field Orders 31,32,33. Were recent additions and shall be completed without delay.

Issued By: LiDestri Foods   IWM3 Mechanical

Site Manager

Director of Business Development

Note that some of the field orders are returned at zero cost or not used. Field Order 17 was simply a transmittal of the torque specification for flange connections. Some orders are made and then cancelled. If we issue multiple orders on a given day, it is easier to keep the cancelled on file and marked not used, as opposed to re-numbering subsequent orders that have been released to the subcontractor.

***

ASIDE

In a similar amount of time, the electrical portion of the job was let at an initial cost of $455,000 and increased to $524,000 after 28 field orders were issued. The process mechanical contract changed from $608,000 to $638,000 over 11 field orders. The total project was delivered on the exact date promised to the customer. Sadly, the customer's customer changed hands right after the job was completed, the projected revenues on the NPV analysis evaporated, and the facility sits idle.

END OF ASIDE

Another best laid plan having gone awry.

CHAPTER 8

More on the nature and intent of documents.

Having discussed the Drawings, Field Order, Request for Information, and Change Order, we briefly turn our attention to a few more documents.

Construction Bulletins
General Orders
Contract Letters
Daily Force Reports
Submittals

ON BULLETINS

A construction bulletin is simply that; a "quick announcement from an official source about an important piece of news" (Merriam Webster). We'll extend that definition within the context of our current chapter by stating… "the construction bulletin is a document that is of general interest to most of or all contractors, the owner, and/or visitors to the site, but not equipment vendors, engineers, Architects etc. that are not generally interested in the day-to-day goings on the job site. Usually, I prefer to issue these by e-mail with a Pdf format document attached, so it can stand out and be printed separately. The bulletin can also substitute and be used interchangeably with meeting minutes. We use the bulletin primarily when the intent of the communication is to advise or caution, as opposed to the General Order, to be discussed later on.

Bulletins simply go out by e-mail, and are collected in our bulletin file. Any bulletin concerning safety gets posted at the entries to the construction zone.

We will discuss further from a library of real examples.

David Glass

Here is a bulletin which announces a temporary disruption to the previously discussed Laydown Plan.

Bulletin Number 7
Gemini Project

Date: January 3, 2013

Issued To: For Information: For Action / Compliance:

☒ MMI   ☒   ☐
☒ IWM3   ☐   ☒
☒ A&B   ☐   ☒
☒ Tetra Pak   ☐   ☒
☒ LiDestri   ☐   ☒
☐ B. Tait   ☐   ☐

This bulletin provides a heads up on some short term activities that affect the overall site and requests some special cooperation in order to keep things flowing. Two substantial deliveries are expected and will require some changes to the laydown plan. Please see the attached for reference.

Monday 1/6:

1. Riggers will remobilize and work with A&B to complete rough-in of Fluidor.
2. Lidestri is requested to remove and dispose of as much surplus furniture etc. as possible along the wall near the receiving door, to make room for filling line cases expected to arrive on Tuesday.
3. A&B to clear as much of the receiving cache as possible by setting panels and relocating Strongarm to the Almix area.

Tuesday 1/7:

1. The pathway to Fluidor is no longer needed, the panels and Strongarm are cleared. IWM3 will relocate laydown to that area. Clearing the area near the roll-up door for Mark the Cooler Guy (MCG)
2. Heavy traffic and some congestion is to be expected as the fill line is brought in.

Wednesday 1/8:
1. Freezer and Cooler panels are expected, filling the MCG area with some overflow into the cache.

If you have any questions regarding this Bulletin, please call Dave Glass at 7245960455, or email to davidg@edsmechanical.com

Here's another one that announces a Power Outage for an electrical service tie-in:

Bulletin Number 9
Gemini Project

Date: January 10, 2014

Issued To: For Information: For Action / Compliance:

☒ MMI   ☒
☒ IWM3   ☒
☒ A&B   ☒
☒ Tetra Pak   ☒
☒ LiDestri   ☒
☒ B. Tait   ☒

There will be a Power Outage next Friday, January 17. It will last all day. Contractors are requested to work on 4-10's or some other work-around.

If you have any questions regarding this Bulletin, please call Dave Glass at 7245960455, or email to davidg@edsmechanical.com

Another Bulletin requests help with locating a pallet of material that misplaced, mislaid, or incorrectly received.

Bulletin Number 6
Gemini Project

Date: December 20, 2013

Issued To:	For Information:	For Action / Compliance:
☒ MMI	☒	☐
☒ IWM3	☒	☐
☒ A&B	☒	☐
☒ Tetra Pak	☒	☐
☒ LiDestri	☒	☐
☒ B. Tait	☒	☐

This bulletin requests a general heads-up / lookout for a shipment that is unaccounted for. The packing list was detached and delivered, but the location of the shipment is unknown. Help is appreciated.

If you have any questions regarding this Bulletin, please call Dave Glass at 7245960455, or email to davidg@edsmechanical.com

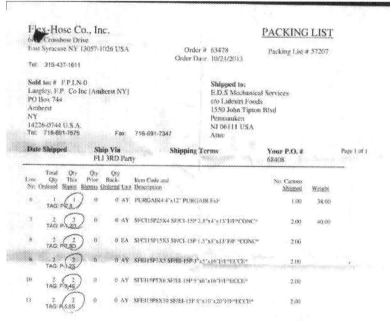

Remember the not-so helpful receiving guy/girl(s) we described back in chapter 5? Let's suppose he received the pallet, left it on the loading dock, and stuck the packing list under our door before he went to lunch. Or worse, we were expecting a rush delivery at heavy expense for guys standing by on Sunday and she moved the pallet to the boneyard before leaving for the week-end. We know these hypothetical bad actors were on duty at the time but there is no direct evidence of evil intent or culpability. The bulletin asks everyone for help in finding the pallet without directly pointing a finger at an individual or department over which we have no direct control. Maybe it was an honest mistake on the part of a new employee, and the more helpful receiving manager gets to the bottom of things before much time is lost after work or on Monday. If this type of bulletin is properly issued and received, we have created a proxy search party – eyes on the ground all over the plant, compared to the construction manager herself, still without the go-to team, searching in the dark corners of the plant with a flashlight.

Next to last, we demonstrate a lengthy Bulletin issued as a record of a construction Kick-off meeting. This meeting is held after the bid award and just prior to contractor's mobilization. The kick-off meeting is a forum where we cement the expectations cast during the bid process and set forth special plant rules. As superintendents and foremen generally do not participate in the bidding, but are charged with running the work in the field, we take the time to emphasize the importance of safety, daily administration, communications, field orders, etc. and get all contractors moving in a common direction.

Bulletin Number 4
Gemini Project

Date: December 9, 2013

Issued To:	For Information:	For Action / Compliance:
☒ MMI	☐	☒
☒ IWM3	☐	☒
☒ A&B	☐	☒
☒ Tetra Pak	☐	☐
☒ LiDestri	☐	☒
☒ B. Tait	☐	☒
☒ Rhoads MW	☐	☒

This bulletin documents the topics discussed and a few we did not address in the Gemini kickoff meeting held at 10 am on Thursday December 5 2013 at the site.

COMMUNICATIONS

Dave Kulak is your Point Of Contact for the LiDestri Facility. Confer with Mr. Kulak (1) On all matters involving permanent modifications to the building eg: wall penetrations, welding to structure roof penetrations, etc.(2) Any use of space outside agreed areas for laydown, work or storage, or, (3) any routing of pipe, conduit or other installed work that may interfere with access or serviceability of existing plant machinery, doors or means of production. Craft supervision is expected to be close enough as to avoid rework around existing plant interferences, and shall not be cause for change by field order whether the route is per approved drawings or not. Dave Kulak shall also be your sole POC for communications with plant employees on the floor. Fraternization between craft and plant personnel should be limited to acts of common courtesy.

Dave Glass is your Point of Contact for all construction activities. There are four main contractors on site as well additional support, and one or two yet to mobilize. Work shall be done in harmony and cooperation. Help each other out when you can. Let someone borrow a scissorlift or forklift for a few minutes if they need to move something or go up and take a measurement and everything else is in use. Daily cooperation among the foremen is essential to everyone's productivity and profit.

Bulletins shall be issued from time to time announcing matters of project wide interest and changing administrative or safety conditions. Communication will be by e-mail to supervisors for further internal distribution.

PLANT SAFETY

Follow the Lidestri safety policy per the slides presented. Beard nets and hair nets required at all times. Fire alarm means evacuate in an orderly fashion through the nearest available exit. Muster for headcount at the Southeast corner of the 1550 John Tipton parking lot. Do not re-enter until an all-clear is given. Plant fire extinguishers shall not be moved or used, except for their intended purpose nor shall they be blocked by materials or unattended equipment.

CONSTRUCTION SAFETY

IAW OSHA at all times, special provisions follow:

1) Fall Protection / Falling Objects

Strict adherence to fall protection is required. Tie off to the handrail inside a lift if there is no joist or suitable structure available overhead. Put caution tape on the floor around overhead work and post a sign requiring hard hats inside the tape. Take special notice of the sloped floor. Small deck platforms are especially susceptible to tipping. Check the operations manual of your lifts and be familiar with any tip-over interlocks. Refresh yourself on the operation and use of the stabilizers. Plan and inspect the path of your lift prior to any moves in an extended position.

2) Hot Work

Hot work can be done under the contractor's own permit system. At a minimum provide 1 20lb ABC extinguisher within 10 feet of any welding, burning or abrasive grinding, fire blankets on the floor. Keep flammables and combustibles 30 feet away. Violations of this policy will lead to stricter measures. Metal on metal drilling, the use of portabands, and sawzalls shall require the same precautions except for the use of fire blankets. Extinguishers shall not be taken up in manlifts unless they are secured in an approved fixture.

Welding screens shall be deployed for all welding operations, on the floor or up in a lift. No exceptions.

3) Use of caution tape / danger tape

Yellow caution tape shall be deployed to delineate areas of temporary, or moving hazards, such as overhead work, rigging operations, wet areas etc. For overhead work post a sign requiring hard hats within the tape, or post a wet floor sign until it is dry. Yellow is the specified color for CAUTION regardless of the printed verbiage. Yellow tape means enter with caution, look up down and around for special conditions and additional warning signs. If you take tape down in order to carry an object across it, put it back up. When you install the tape, tie it with a bow so it can be put back up easily. Leave tape up overnight if necessary. Remove and dispose of the tape when the hazardous work is completed.

Red tape shall mean "DANGER, DO NOT ENTER", except on authorized invitation. Use red tape when there is a clear, present and unavoidable temporary hazard, such as an open energized panel, a hot tap, leading edge work, open vertical manways or floor openings, or moving machinery which is undergoing testing without permanent operational guards, interlocks, or safety fence. Remove and dispose of the tape when the hazardous work is completed.

Do not use one color tape in place of the other. In either case, for very short durations, you may post a dedicated sentry who is familiar with the hazard and maintains line of sight between the hazard and anyone who may approach.

The door to the rail shed at Column 3A requires special consideration. Civil work is taking place immediately outside and conditions are changing rapidly. Grade elevation will be changing and trip hazards and obstructions will exist. It is easy to get locked out. A new roof is to be installed and electricians will be putting new breakers in the Main Power Panel.

4) Lock-out / Tagout

Lock-out and Tagout can be a difficult procedure to define and implement with multiple contractors, each having different policies, inside an operating plant.

It is assumed that general training has already taken place and situations requiring lockout are self-evident to the trained worker. Identification of the sources of stored energy, analysis of branch circuits and other energy paths, confirmation of and zero-energy states are not addressed here.

Areas of general acceptance:

Locks may be placed or removed at any level of management to the level of trained apprentice; or controlled at a higher, more centralized level, such as crew foreman or general foreman.

Locks may be keyed alike or uniquely. If a centralized, keyed alike system is desired, there shall be only one keyholder.

Locks shall remain in place until the work requiring the lockout has been completed, and shall be removed promptly thereafter.

The use of commercially available LOTO devices for unusual shapes (breakers, valves) are encouraged. Alternative devices may be approved by the construction manager.

Special requirements for Gemini Project:

The first device to be attached at the lock-out point shall be a lockout hasp capable of receiving additional locks as required. The last available hole in the hasp shall receive another hasp, which will enable the placement of additional locks. And so on.

The person placing the lock shall identify it by means or sturdy, commercially available tag, affixed to the shackle of the lock with a strong nylon tie-wrap. The name of the person placing the lock, company affiliation, company phone number and person's cell phone number shall be clearly legible on the tag. Non water soluble ink shall be used, Sharpie brand pens or equal work well.

No one shall be authorized to remove a lock placed by another person. This may be considered a criminal act of reckless endangerment or worse depending on the circumstances and outcome.

OUTAGES

Superintendents and project engineers should begin to examine their scope of work for outage requirements – interrupted services, dates and durations. An outage plan shall be developed with the plant and construction manager to minimize disruptions and coordinate the work. Check the mechanical and plumbing services for valve locations that may be closed in order to avoid a wider outage than required for the work to be done.

One outage at the main power panel has already been requested to install new breakers, and may also be used to megger the main feeds from the utility transformer. We have requested this to be done before Christmas, and will communicate the time and date by Bulletin.

INGRESS and EGRESS

Through the turnstile entrance only. Other doors shall not be left or blocked open unless they are attended by line of sight. Please note the double doorway in the receiving area. There is a potential to get locked in the vestibule or outside the building.

PROX CARDS / BADGES

One per contractor or entity as approved. Badge list with name and phone number shall be maintained by the construction manager. Issuance of a prox card means YOU have been trusted with unrestricted access to the facility. Cards are not to be passed around for unscheduled smoke breaks or transferred to another supervisor without approval. Cards are the property of LiDestri and may be recalled without cause or notice. Treat it as you would a credit card or driver's license.

HOUSEKEEPING

Trash in receptacles, not on the floor, receptacles on pallets or wheels so they can forked or moved without dragging across the floor. C&D dumpsters are provided by LiDestri, let Dave Kulak know when they are getting full. If your craft is seen overloading a roll-off they will be required to return the whole load to an acceptable level. Work areas shall be broom swept at the end of each shift. This includes the decks of man lifts. Work in cooperation on common area end of shift clean-up.

Salvage or disposal of crating or dunnage is at the sole discretion of Dave Kulak. Contractors shall keep title to any scrap metal under their lump-sum scope, Lidestri shall keep title to any scrap paid for on time and materials.

LAYDOWNS and AISLEWAYS

Refer to the map provided with Bulletin 1. Laydowns were established based on reasonable expectations prior to contractor's mobilization. If space runs short notify Dave & Dave and we will make additional accommodations. Do not just claim space by taking. Your work is taking place in in an operating plant under changing conditions. The designated laydowns are for the Gemini project only. There is another hot-fill project working concurrently by others as well as ongoing production. The designated aisleway is considered a path to safety and shall remain clear at all times.

CONTRACTOR'S EQUIPMENT

.

Oil leaks are prohibited. Leaking machines shall be immediately removed from service and work stopped until a replacement is delivered. Temporary containment fashioned out of sorbent pads, duct tape, plastic sheets etc. are prohibited. Deliveries of bulk propane shall be made at least 30 feet away from any building or source of ignition. Full bottles shall be stored outside in a locked cage, or off site. Bottles may remain on equipment inside overnight with the valves closed. No idling of delivery vehicles inside the building.

Keys shall remain in equipment parked inside overnight and on weekends, otherwise equipment shall be parked outside in an area designated by Dave Kulak.

Electric equipment should be parked in the designated laydowns if receptacles exist, or along the walls. Use the same location for charging each night. If there are not enough receptacles, work together and share a pigtail. Anyone discourteous enough to pull a plug will be removed from the job.

USE OF LIDESTRI FACILITIES

As authorized by Dave Kulak.

MSDS SHEETS

Submit as required for all hazardous chemicals on site. Send an e-mail link to the sheet to davidg@edsmechanical.com I will maintain the list and manage further distribution. Do not send any links that do not point to the MSDS sheet with more than one click. If a direct link cannot be found, submit on paper.

FIELD ORDERS

Field orders are issued to authorize any and all changes in the work. Do not do any work outside your contract and expect to get paid unless you have one. Field Orders may also be issued to document engineering changes that do not result in a cost change, or result in a credit.

DAILY FORCE REPORTS

Use the form provided, return one for each day craft is on the site. Keep separate time against all field orders. For example, if a man works four hours on a field order and four on contract return two force reports for that work, 4 hours each. Force reports are due 10 am on the next working day.

Finally, here is a Bulletin which captured the overall sense of cooperation and goodwill we sought to establish on this project, when a Journeyman had the courtesy to turn in a wad of money which was found on the floor along with a credit card receipt from the local diner:

Bulletin Number 8

Gemini Project

Date: January 10, 2013

Issued To:	For Information:	For Action / Compliance:
☒ MMI	☒	☐
☒ IWM3	☒	☐
☒ A&B	☒	☐
☒ Tetra Pak	☒	☐
☒ LiDestri	☒	☐
☒ B. Tait	☒	☐

A few dollars were found on the floor of the work area yesterday along with a credit card receipt for lunch at the Penn Queen Diner.

The Money shall be returned to the bearer of a VISA card ending in 1878 upon display to the Construction Manager.

If you have any questions regarding this Bulletin, please call Dave Glass at 7245960455, or email to davidg@edsmechanical.com

And the fellow who lost his money got it back just like that.

Now, the foregoing examples on the appropriate use of the bulletin have been more or less demonstrated and we close discussion on the document by noting that one of the unwritten intents is that the bulletin be issued in the form of an *instruction,* i.e.: a one way communication not for further comment or reply, unless there is a reason within the document like wanting to return lost property.

GENERAL ORDERS:

We now take a moment to briefly discuss the concept of the General Order by referring to the definition of the two words (credit Merriam Webster)

General: involving, applicable to, or affecting the whole
Order: to give an order to: command

A General Order is like a bulletin, but we use it only occasionally when the issue affects everyone on the entire site, like a change in the general conditions, or change in general policy, or the one or two items we did not plan for. General Orders should be kept to one or two sentences. We keep them numbered sequentially, use them as sparsely as possible, and maintain a field book in the construction office for quick perusal by visitors. The General Order gives us a good firewall against acts of insubordination.

Some examples:

- Due to the abuse of smoking policy, further use of tobacco in any form whatsoever is prohibited and shall be cause for removal of the offender upon first offence.
- Persistent noise levels over 90 Db have been recorded recently by the Project Safety Engineer in the Pump Room. Hearing protection shall be required upon entry.
- Due to the increased levels of craft on the site, contractor's superintendents shall be required to attend a morning coordination meeting, daily between 0715 and 0730 at the office of the Construction Manager.
- The Lab Room construction has been accepted and turned over to the owner. Further entrance by contractors is prohibited except by special permission.

CONTRACT LETTERS

Contract Letters serve two main purposes:

Routinely, contract letters are used when the parties established a requirement for written notice under a term in the contract, such as a:

Notice of Violation
Notice of Non-Compliance
Notice of Intent to Terminate
Notice of Breach
Notice of Change

Notice to Proceed
Notice of Substantial Completion

If such an event occurs, the contract letter may be initiated by either party. For legal purposes, we may or may not have written a requirement that they are transmitted by registered mail, return receipt requested. The first four kinds of letters probably indicate the unfortunate situation where a problem could not be reconciled amicably in the field, and the issue has been kicked up the stairs to the management suite. The letters document the arguments on each side of the issue, and create a record of escalation, to senior management, the executive, arbitration or legal action.

The Notice of Change, or Change Order Request aka (COR) usually originates from the aforementioned change driven contractor, or a nobler contractor who has routinely been denied field orders by a stingy owner, or when omissions or issues raised in the bid process were left unresolved. If the day to day change issues and small problems are not resolved as they arise in the field, they will escalate, time will pass, the facts forgotten, and the money finally closing the issue will have little correlation to the actual expense in the favor of the subcontractor or owner. To avoid the situation in this paragraph and remain on good terms, we refer again to axiom 2 – 65% planning. Another good idea is to place sufficient trust in the field. The owner's rep or construction manager should have final authority to resolve single issue changes valued up-to 1% of the contract value e.g.: $20,000 on $2,000,000 or draw from the contingency budget as she sees fit.

The Notice to Proceed, NTP (sometimes Letter of Intent) is a document used in cases where there is a corporate bureaucracy that may take a substantial amount of time to formally approve the contract. This usually happens in larger corporations where a contract may be prepared by a project manager or procurement but can only be signed by a corporate executive or officer. Funds have been authorized, the project approved, but the paper! The Paper! The NTP tells the contractor to get going and guarantees payment of costs incurred before the contract is finalized.

ASIDE

I have worked several projects in a Japanese context valued in the $50 - $250 million range where the contract was executed after the job was done. The only document in force as the work got done being a one or two sentence letter of intent. Within the Japanese culture a development trust between the parties, the bow and the handshake are of paramount importance; paper more suited to the fine arts such as Calligraphy or Origami.

END OF ASIDE

NOTICE OF SUBSTANTIAL COMPLETION

This is a contract letter generated by the contractor informing the owner that the project is substantially complete and the time has come for the owner to turn his efforts to punchlist generation, acceptance, turnover, and final payment. These four shall be discussed in the chapter titled Closeout.

SUBMITTALS

Submittals are documents or even physical specimens required of the contractor for approval by the owner, construction manager or QA/QC manager. The submittals requirement depends on the complexity and risk of the job. We might not require any for the construction of a doghouse up-to including everything imaginable for a nuclear project or passenger jet aircraft. You can request a submittal on just about anything imaginable. Just keep in mind that some submittals are cheap or cost next to nothing and some can be very expensive.

A garden variety list follows:

Material Submittals

      Engineered Fill
            Certificate of Origin
            Certificate of Compliance with ASHRAE
            Chart of Particle Size Distribution
            Physical Specimen

Unclassified Fill
>   Dry Density
>   % Moisture
>   In Place % Compaction
>   In Place Nuclear Density

Concrete
>   Results of Slump Test
>   Results of 7 and 28 day Compressive Strength Tests

Metals
>   Certificate of Composition
>   Certificate of Moh's Hardness
>   Certificate of Rockwell Hardness
>   Certificate of ANSI compliance
>   Heat Numbers

Lumber
>   Certificate of Origin
>   Chemical analysis for pesticides or undesirable chemistry
>   Percent Moisture

Fasteners
>   Certificates of Non-Destructive or Destructive Testing.
>   Verification of ASME grade

Purchased Parts and Machinery
>   Spec Sheets, Data Sheets, Performance Curves
>   Operations Manuals
>   Recommended Spare Parts List
>   Maintenance Manuals
>   UL Label

Instruments
>   Same as above with the inclusion of Calibration Data, Accuracy, Resolution, and Precision.

Engineering
> Drawings Stamped by a Professional Engineer
> Rigging Plans
> Engineering Calculations
> Letters of Professional Opinion
> Erosion Control Plan
> Rainwater Runoff Control Plan

Construction
> Daily Force Reports
> Torque Specifications
> As-Built Drawings
> Certificates and Requests for Payment
> Weld Maps
> Results of Weld X-ray Testing
> Results of Hydrostatic Tests
> Results of Pneumatic Test Results

Humans
> Criminal Background Checks
> Drug Test Results
> Identity
> I-9 Status
> Resumes
> References
> Diplomas/Professional Certifications/Professional Society Licenses
> Security Clearance
> Certificate of Radiological Health (Department of Energy)

There are several reasons for requiring submittals:

- For equipment - pumps, tanks, compressors, etc. and architectural finishes, we have a paper chain from the RFQ to the packing list. What we wanted, to what finally got installed on site.

- If the specification allows material or finishing as "Or Equal" or "Equal or Better", we have a written record of the comparison, and who approved it.

- The construction submittals taken together help to ensure the overall quality of the job.

- The engineering submittals are usually needed somewhere along the line for government approval and/or permitting.

- The human submittals have been listed for reference and not really our business for the most part. We might ask for the project manager's resume and references, and we would need clearances on any one in a classified or public trust area. We assume the contractor has done a criminal background check as a matter of policy, and the I-9 requirements have been met. We clarify these assumptions during the award conversations and fill in the gaps with the successful contractor rather than asking submittal as part of the bid.

And now, the attentive reader may have noted that we have not discussed the Daily Force Report in any detail. Is that because it is a contractor's internal use document? Not our business because he is working on a lump sum? The contractor might very well make that assertion, and that's OK. We can always use someone else at this point. The timely receipt (next day at 0900) of a Daily Force Report, henceforth DFR, containing the information we specify shall be a contract requirement. The DFR is of such unique and singular importance that we might even write into the contract language such as:

Owner or Construction Manager may order a stoppage of all work if the contractor falls 5 days behind submittal of DFRs. Stoppage shall be at the sole cost of the contractor, the end date shall not be extended.

Owner may fine the contractor $1 for each hour the DFR is delinquent.

The foregoing, submitted more or less tongue in cheek represents one of the most systemic problems I have experienced over 30 years, is the subcontractor's frustration in receiving *their own* DFR's in order to handle payroll for *their own* employees, and now you want the same information on your own form, delivered each and every day?

WOWZERS.

Let's turn the page, since the daily force report merits more focused attention in the next chapter.

CHAPTER 9

Elements of Construction:
Bricks, Mortar, Steel.

The Daily Force Report (DFR) is Another.

We have on more than one occasion used a part of the (5 W's) - Who, What, When, Where, Why, and sometimes (6) How, or (7) by what means, as a lever to get the whole story, or get to the truth. The record set of drawings is one, the RFI's chain another, contract correspondence letters a third. This device, widely used in the first paragraph of every newspaper report, can be traced to the time of Hermagoras of Temnos (Greek. 100 b.c.) and has been used every day in one form or another ever since.

Our DFR is elemental, like a brick in a wall, or rebar in concrete. For each worker on the site, we ask Who (Bob), What (Pipefitter, Journeyman) When (Today, 8hrs), and Where (SOV Line # or FO#). The requirement is across all contractors on site, let's say 5 subs, 20 workers each, and suddenly we have 100 Bobs, 8 hours a day, 800 hours, 5 days a week, 4000 hours total or, at the rate of $77 per hour, $308,000 per week in labor. The DFR is our record of how every single one of those dollars were spent, and allows us to know how much each task took to complete down to the hour or half hour if desired. In addition to the 5W's, we allow a space on the DFR for the foreman (or timekeeper) who prepares the report to make note of any unusual events or circumstances that might impede the work and require attention from the owner or CM.

Now recall how we took the trouble to prepare a bid form, itemizing in an organized fashion, the discrete tasks, or scopes of work. Then we asked the contractor to submit this form as a schedule of values, with his bid, and carry the effort into the time domain by preparing a man-loaded schedule in accordance with schedule of values.

Why did we go to all this trouble and then ask for more?

The Bid form, schedule of values and man-loaded schedule have paved our road, and set the course for the job from end to end. Now we close the ring by constructing a string of DFRs that metaphorically is our truckers log or ships log, with each hour recorded against an SOV or MPL line item. Just like a report on the position of a ship at sea, or more recently, the position of anything that might be tracked with an RF-ID tag.

Let's imagine the hypothetical job we have bid and scheduled has been completed some time ago. For the moment I'll assert some artistic license for illustration. We pull a DFR out of the job file at random and it looks like this:

Industrial Piping Co, Inc.												
Uniform Daily Force Report												
Gemini Project												
Date	1/8/14											
		SOV		SOV		SOV		SOV				
Name	Craft	Line	Hrs	Line	Hrs	Line	Hrs	Line	Hrs	F0#	Hrs	Total
Larry Rose	PFM	1	2	16	2	7	2	10	2			8
Jay Dougherty	PFJ	1	8									8
Jim Jalkowski	PFJ	1	8									8
Fred Weidenmiller Jr.	PFJ	1	8									8
Fred Weidenmiller Sr.	PFA	1	8									8
Pat Leister	PFJ	16	8									8
Aaron kalvitis	PFJ	16	8									8
Dan Aubry	PFJ	16	8									8
Mike Casale	PFJ	16	8									8
Len Juliani	PFJ	7	8									8
Joe Brunner	PFJ	7	8									8
Tony Bendetto	PFJ	7	8									8
Chris Gonzalez	PFJ	7	8									8
Paul Kyrzna	PFA	7	8									8
Brad Poeth	PFA	10	8									8
Joe Nestor	PFJ	10	8									8
Dino Amick	PFJ	10	8									8
Dave D'Appolonia	PFJ	10	8									8
Simon Swedburg	PFJ	13	8									8
Brandon Cole	PFJ	13	8									8
Steve Hecknauer	PFJ	13	8									8
Larry Speer	PFJ	13	8									8
Total	22											176

Visitors:

Name / Business
None

Problems or Unusual Conditions:

Prepared by:			Filename Protocol - Save as
Larry Rose			COMPANYNAME_DFR_MMDDYY.xlsx

For the sake of argument, let's suppose the job started on January 1ˢᵗ, and for the sake of simplicity and not knowing beforehand when the job would start, we constructed our MPL timescale from a hypothetical day 0 forward. We take January 8 on our DFR = Day 8 on the MPL.

Now, we take these documents side by side and observe:

MPL (Planned) January 8   DFR (Actual) January 8

22 Fitters on site 176 Hrs Total   22 Fitters on site 176 Hrs Total

4 Fitters working SOV line 1   3 Journeymen, 1 Apprentice working SOV line 1

4 Fitters working SOV line 16   4 Journeymen working SOV line 16

6 Fitters working SOV line 7   4 Journeymen, 1 Apprentice working SOV line 7

4 Fitter working SOV line 10   4 Journeymen working SOV line 10

4 Fitters working SOV line 13   4 Journeyman working SOV line 13

Note 6. Productive labor only   1 Foreman SOV lines 1, 6, 7, 10.

And now, we note that the work we planned, for this day (MPL), all the way back when the job was bid, looks like it actually happened, for this particular day at least. We should be very happy.

But wait…Not just yet.

This comparison shows only the planned labor vs the actual for a single day (Hours Planned vs Hours Actual). We must now examine what we should expect have accomplished on this day.

Recall back in Chapter 5, part 2 we took the bid form and calculated expected production rates from the linear footage/budgeted hour to come up with expected production rates for each type of material. Repeated here, we have:

SOV line 1
    Carbon Steel Header .76 feet per hour.

SOV line 16
    Steam and Condensate 1.4 ft/hr

SOV Line 7

> Pump Room 427' = .20 ft/hr for piping.

SOV Line 10

> Compressed Air 1.5 ft/hr

SOV Line 13

> City Water 2.5 ft/hr

Let's continue with this oversimplification for the sake of illustration. On January 8, with the crew as reported on the DFR working unencumbered, on relatively straightforward work, at the estimated production rates, we will install:

SOV line 1 – 32 hrs x .76 ft/hr = 24ft
SOV line 16 -32 hrs x 1.4 ft/hr = 45ft
SOV line 7 – 40 hrs x .20 ft/hr = 8 ft
SOV line 10 -32 hrs x 1.5 ft/hr = 48 ft
SOV line 13 – 32 hrs x 2.5 ft/hr = 80 ft

Now, we start to free ourselves from the bonds of oversimplification by asking if we go to work in the morning on January 8 and make an observation as to progress, can we expect to go walk the site at the end of the day and have this much new work completed? After all, we had gone to such great lengths to prepare our bid package, instruct the bidders to show hours and quantities on a schedule of values, required a man loaded schedule, vetted the bidders thoroughly to avoid miscreants and scalawags, set the mutual expectations to the highest bar, cleared the aisles and got out of the way.

Still, the answer is maybe yes, but probably no.

The reason why we cannot answer yes with a higher degree of confidence is not due to any flaw in our planning, scheduling, or preparation. It is simply the window of observation (8 days) is too short to be of use. The production rates we have presented are an aggregate of the whole period of construction. At any given point in time, our actual rate may very well be zero (ran out of material, did something else), as scheduled, or maybe faster than expected. Picture a typical trip down a random interstate highway. The speed limit or maximum allowable rate is 70 mph. There may be periods of congestion where we slow to 45 mph for a time, or an unscheduled stoppage (flat tire) 0 mph for a short time, or an unexpected detour, going the wrong way altogether until we pass an obstacle (egad! rework!). Then we have one of those days when the sun is out, there's nobody on the road, the police are at the donut shop and we can set the cruise on 85. Where we are at any point in time, especially in the early stages of the job or trip, may not be immediately evident.

There are also "end effects" such as work that has to be done at the beginning or end of the scope but not, for the sake of simplicity, indicated as a separate task on the MPL. For piping, we have to fabricate and install hangers or other supports before pipe can be run. At the end there can be extensive flushing and pressure testing. For rigging work, there might be an extensive inspection before the machinery can be removed from the truck. For precisely leveled and aligned equipment, there will likely be a manufacturer's verification of the millwrights work.

This idea of planning our trip at 60 miles an hour overall, from point to point, is exactly the same as planning our production rate for the city water at 2.5 ft/hr. From one end to the other, New York to Los Angeles, boiler to condensate drain, we can expect to get it very close.

The metric we chose for measurement (linear feet) is the best one available, but we have a few bumps in the road. There are elbows, hangers, valves etc. in the path, items which take time and don't count for too many lineal feet of installed work. If we want to get more precise, we can, but we run the risk of turning the job into a measurement science project and losing track of the big picture of progress. Don't forget, our piping scope is only one of several others going on at the same time. At the same time we are tracking the electricians, tin-knockers and rodbusters.

As a final thought, we might have trouble with the actual definition of "Installed." Some assemblies of work might very well be prepared out of mind's eye in the shop or elsewhere off – site, and shipped ready to be installed in the blink of an eye. We do not immediately complain if we don't see something getting done. The work might just be sitting on the truck outside. That's another reason why we have our daily 15 minute meeting.

For the present case, the best thing to do to more accurately quantify the progress for the work at hand is to widen the observation period another week and repeat the analysis. The DFR's continue to provide a discrete and actual measure of the labor consumption per SOV line, (Truth) and the materials can be progressed as desired.

The more experienced reader might surmise at this point that we are heading into a discussion of Earned Value Management, and yes, we are. We will, however refer the curious to a DOE website for an entry level discussion on the subject: http://energy.gov/em/services/program-management/project-management/earned-value-management, or other sources of existing literature for the academic discussion where the planned value (PV) and earned value (EV) curves are handed over to your consideration in the abstract. We will proceed to *build* our curves keeping in mind the principle that we can construct anything we want and derive understanding from the basic elements of mass, length, and time, as we discussed in chapter 2. We will stick to our elements (mass, length, time) using pounds, feet and hours. Realizing that a man-hour (a unit of power) can be derived from the 3 elements, as can a dollar.

For example:

If one horsepower = 550 (ft x lbs)/second, we can safely say that in the aggregate, 1 manpower = one horsepower x some constant C. C, is unique for each worker ranging on a scale from someone like Calista Flockhart ($C=0.15$) on the low end, Arnold Schwarzenegger in the middle ($C=0.5$), and one of those Budweiser horses ($C=1.2$) on the far end of the scale.

Time *is* money, and in our case it is exactly $\$1 - 1/$(labor rate)/man. Or for one guy or gal working at $75 per hour:

$\$1 = 60/75 = 48$ seconds of his or her undivided attention.

ASIDE

And now I beg the forgiveness of my readers. Heretofore we have gone to considerable length to avoid any subjective differences between the sexes. In fact, when comparing the differences between Arnold Schwarzenegger, Calista Flockhart, and a random horse of either sex, we have reduced these differences to a simple number we shall call a "Power Factor". Henceforth we shall call a day's labor a man-day, or a labor hour a man-hour, and human output man-power. This is for brevity only with all due respect to the other

196

sex. In fact there are only three paid jobs in all the world which are gender specific. Can you guess what they are?[1]

In any event, going forward, a man-lift is a man-lift and we don't care who gets in one, as long as it is not some type of horse.

END OF ASIDE

---

[1] Sperm Donor, Wet Nurse, Surrogate Mother

Chapter 9 Part 2. Constructing the earned value curves

Before we begin, let's review the basic elements of a line graph. Don't worry, you did this in high school, and see them every day in one form or another. The graph has an origin, a horizontal X axis, and a vertical Y axis. The origin is usually at the intersection of X=0 and Y=0, and usually in the lower left corner of the graph. For the purposes of an EV chart, the X axis represents the passage of time, from start to completion, and the Y axis another unit of measure, which represents the accumulation of resources consumed, (man-hours or dollars). The Y axis can have additional scales of concern – 0-100% complete, or 0-total linear feet, tons, installed sq-feet, etc. anything we wish to track and control against time. Now, we keep in mind that we begin with a chart that starts at mobilization lower left (0,0) and ends at the upper right (scheduled completion, total budget).

If the foregoing paragraph has set a ball of confusion in motion, don't worry. Unlike high school math, our line charts only exist where x and y are positive. (time and spending only move in a positive direction) So we confine our concern to the upper right quadrant of the line chart as a range of possibilities. Another way of saying that we only expect the reader to recall 25 % of what the math teacher was trying to hammer home in 10th grade.

We can, and will, construct an EV curve for our demonstration project MPL, and note that for an additional exercise, the reader can construct a similar graph for each of the SOV items discussed in part 1 of this chapter.

Let's get started. Most of the work is already done. Our MPL starts at day 0 and goes to day 55. This goes on our x-axis. The y-axis data consists of the total man-power predicted and accumulated at the end of the day. This is the total of the previous days plus the current day.

Logically, we can write Total (today, COB) = Total (thru yesterday) + work done (today)

Or in other words, at any point in time:

$$T_{i(COB)} = \sum_{(i=0)}^{n} T_{(i-1)} + T_i$$

Or, in excel, our data table emerges and starts out looking like this, truncated after 14 days for brevity:

Days From Start	Workers on Site	Man-days accumul.	Man-hours accumul.
0	3	3	24
1	19	22	176
2	19	41	328
3	19	60	480
4	19	79	632
5	16	95	760
6	22	117	936
7	22	139	1112
8	22	161	1288
9	22	183	1464
10	22	205	1640
11	22	227	1816
12	22	249	1992
13	22	271	2168
14	22	293	2344

Graphically, we can draw the curve for whole project from start to finish like this:

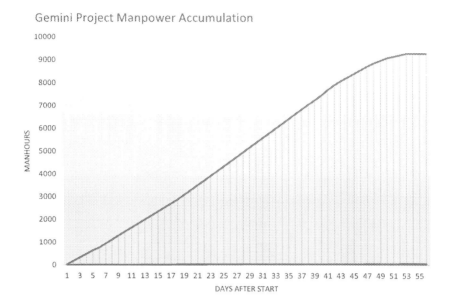

Gemini Project Manpower Accumulation

We have our projected EV curve for the whole Job. As noted previously, time is money, and the vertical axis might as well be expressed in dollars. $61=1hr. This line represents where we want to be at times, and does not change.

Gemini Project Manpower Accumulation

Now, we start to take our DFR's as they come and we plot the actual hours worked, every day. Our bold black line will be our actual curve. We started 5 days late, and our sub only managed to work 1000 hours in the first 15 days, 1520 hours short of the 2520 planned. Smells like trouble, but can we see where the trouble lies? If we refer back to our bridge analogy in chapter 5 part 2, we have suffered both a late start and, once out of the gate, we stumbled. Applying simple math to the situation at the close of day 15, we have:

40 days remaining, 9272-1000 = 8272 hrs to go = 206 hrs/day

206/8 = 25.75 men/day. This is above the 21.12 planned going forward. And those three days of zero labor we held in reserve at the end are gone just like that.

AXIOM 17

IF YOU HAVE NOT UPDATED YOUR EV CURVES IN THE LAST THREE DAYS, YOU ARE LOSING CONTROL OF YOUR PROJECT

Now, in a friendlier world, our situation might very well look like this:

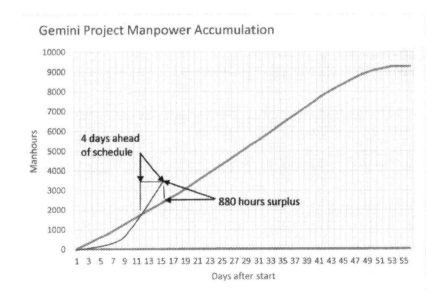

Here our sub has worked 3400 hours, ahead of the 2520 planned through day 15. He seems four days ahead of schedule, or is 3400-2520 = 880 hours or $61/hr - $53,680 ahead on budget.

And a rose by any other name would smell as sweet.......right?

The trouble in either case is that our view remains confined to the abstract. Planned hours vs. actual hours. To get the whole picture, we must look what we expect to have *accomplished* during this period of time. Again, (we did this once already) this information is already before us on the manpower loaded schedule.

Our planned hours through day 15 by SOV line … (Rates per Ch 5, Part 2)	Expected Result
SOV line 6 100% complete 88 hours consumed	8 pumps set, chiller, tanks H1 set
SOV line 9 32/110 hours = 29% complete	Cool tower set on roof, pipe started
SOV line 1 480/1743 hours = 27.5% complete	CS piping .76 ft/hr - 364 ft installed
SOV line 16, 18 480/853 hours =56.3% complete	Steam/Cond. 1.4 ft/hr -672 ft installed
SOV line 7 480/2134 hours = 22.4% complete	Pump Room .20 ft/hr -96 ft installed
SOV line 10 100% complete 448 hours consumed	Line 10 CA 100% complete
SOV line 11 4/340 hours 1% complete	Maybe some hangers installed
SOV line 13 100% complete 384 hours consumed	Line 13 CW 100% complete
SOV line 14 96/662 hours 14.5% complete	CW 1.1 ft/hr -728 ft installed

Now having calculated the expected result from our SOV and MPL, we make a special walkabout and measure the actual results on Day 15. Imagine, low and behold, that the expectations are in line with the actual installed work. Referring to figure 1, we may conclude that the job is going better than expected for the contractor. Instead of raising a warning flag that he must add manpower, the truth is that he will likely complete the job on time and under budget.

In figure 2, if our contractor has consumed 3400 hours as of day 15 and got the same amount of work installed, he is in big trouble. In the first case we spent 1000 hrs x $61 per hour =$61,000; in the second, we spent 3400 hrs x $61 = $207,400 to get to the same place.

Of course, these two charts have been presented to illustrate what might be rather extreme deviations from the expected. Since we have obeyed axiom 2 with fervor, and having parsed out our scope into such discreet packets, it is unlikely that we will stray more than five percent from our projected line unless there is a hidden type 1 error.

Now, the thing to note is that the MPL, SOV, and EV curve have been prepared well ahead of mobilization. The amount of time it takes to add the actual data at the end of the day is trivial, more so if we enjoy the help of a project assistant. We apply the data to the curve every day from the start and stick to it.

For a greater degree of control, we could, and should, build an EV curve for each SOV line. We have built the DFR to accommodate one Journeyman working on as many as 4 different SOV lines and one field order on a given day. We can handle the data, so we might as well use it. If we create an EV curve for each SOV line, we will glean more detailed knowledge as to exactly what is ahead of schedule and what lags behind. I'll leave construction of these curves to the enthusiastic reader as an exercise.

In the real world, one of the most difficult and protracted problems faced by contractors is getting the foremen to turn in contractor's own time, even on a weekly deadline in order to meet payroll. From this point of view, there may be some initial resistance to using paper dictated by the owner or construction manager. Some contractors use boilerplate timesheets that are completed on a weekly basis. In this case we ask the foreman to come to the office and scan the data daily. If the job runs more like six months or a year, we might get away with adjusting the curves on a weekly basis. The job is yours to manage.

Using the EV curve as a forecasting tool:

Now that we have demonstrated that we can accurately measure the installed work because we constructed discreet and quantifiable SOV lines, let's take a closer look at Figure 1 and our EV curve. On day 15 our contractor has worked 1000 hours and has completed what was estimated to take 2520 hours. The slope of our actual curve is 1000hours/10 days, or about 100 hours per day. The projected slope was 2520hours/15 days or 168 hours per day. Since the installed work is precisely on schedule, we can take a straight edge, anchored at the start date (t+5), and apply the slope to day 55. Returning to the 10th grade, we have, the equation for a straight line:

Y= mx +b

Let b=0 at day 5 (translate the origin, since nothing happened during those five days)

m=100 hrs/day

x=50 days

On Completion Y=5000 hours

Time is money: $61/hr x 5000 hrs = $305,000

Recalling our SOV, the dollar estimate for labor was $562, 070. It appears our sub is enjoying a nice job. If those guys in the golf cart are paying any attention at all, they are going to come down to the field office one day and ask you "Can you explain to me, once again, why we didn't do this job on T&M?"

And of course, being the prudent CM, you made a detailed note of who, why, where and when that decision was made.

Now, let's look at Figure 2 and repeat the exercise. Our slope is 3400 hours in 15 days or 226 hours per day, the installed work is the same as in the first case, so we can say:

Y=mx+b

b=0, since the work started on time,

m=226
x=50 days

On completion, Y= 11300 hours
Time is still money $61 per hour x 11300 = $689,300

A very bad job for our contractor. And those guys in the golf cart pat themselves on the back and say, "Sure is a good thing we did this on a lump sum! Heck!, might as well play another round.

Two end notes on forecasting: First, think of the typical forecast you see on the weather channel for the path of a hurricane. There is a best guess line, which is constructed from the best measurable information available using the best computer algorithms available. Still, the forecasters also present a cone of uncertainty. The present situation is known, (a single point on the weather-map) but the predicted location of the storm is less and less certain the farther we look into the future. The nature of a construction progress forecast is similar. In the foregoing example, we looked at information through day 15 and projected to day 55. As time goes by, more and more work is done, the costs known and the forecasted period gets shorter. Just as with the weather forecast, *We can only be more and more accurate with the forecast as time passes.* The trouble lies with management that expects the forecast to be unchanging.

Second, for the sake of simplification, when we constructed our slopes, we took our sample from day zero to day 15 and calculated a straight-line average. This is fine if the hours per day are relatively consistent across the sample period (the usual case). If the data is scattered we calculate our slope using the root mean square (RMS) of the data. This is simply the square root of the mean of the sum of squared data.

Or:

$$x_{\mathrm{rms}} = \sqrt{\frac{1}{n}\left(x_1^2 + x_2^2 + \cdots + x_n^2\right)}.$$

This is also known as the "best fit line".

Before we close the loop on the uses of the schedule of values, manloaded schedule, and daily force reports, we take a moment to take the guess work out of invoicing and payments. Earlier, we suggested that if the preparation of the schedule of values were left to the sole discretion of the contractor, we could expect that he or she would front load costs to the extent allowable. Lots of money for mobilization and work scheduled to come early, and less later. The schedule of values in this case having nothing to do with true value earned, and more to do with grabbing sufficient cash to self-finance the project. By specifying the format and line content of the schedule of values, we better match the contractors revenues to his expenses, and our cash outlays matched to the value of the work installed. We reproduce our schedule from chapter five for illustration.

Process Utilities Scope of Work -Bid Form

Line Item	Area	Drawing	Scope	Reference Documents	Total Linear Feet	Labor Hours	Labor	Materials	Equipment & Consumables	OH&P	Total
1	TA 1 and 2	8	6", 4"and 1-1/4" from Pump Room wall near D9 to TA 1&2 plus 16 drops. Continue header to CL 4.		1297	1743	$102,316	$29,269	$4,400	$17,200	$153,185
2	TA 1 and 2	M1.11	Insulation for Glycol Piping above		807	198	$17,480	$6,927	$1,200	$640	$26,247
3	FCU1	M1.12	Continuation of 1-1/4" from CL4 to FCU in MCC Room		125	115	$8,200	$616	$1,300	$1,500	$11,856
4	FCU1	M1.12	Insulation for Glycol Piping above		186	65	$5,880	$1,658	$560	$450	$8,548
5	FCU1	M1.12	Installation of Trane FCU and Ductwork in MCC room	FCU 1 15 2013 (MS Word), M1.15	NA	64	$4,352	$1,140	$450	$325	$6,267
6	Pump Room	M1.13	Receive and set P1-P8, Chiller, Glycol Tank, Tower Water Tank, and HX1	LiDestri Foods Process Chillers (MS Word), Pump HX submittal (pdf)	NA	84	$5,712	$4,337	$575	$356	$10,980
7	Pump Room	M1.13	All piping in Pump Room, and Between Room and Chiller. **Submit isometric for approval.**		427	2134	$120,682	$66,624	$3,825	$23,144	$214,275
8	Pump Room	M1.13	Insulation for above		255	134	$11,760	$3,934	$553	$296	$16,543
9	Roof	M1.14	Receive and set Cooling tower on existing support steel. **Submit transition piece drawing for approval,** fabricate and install transition. Complete tower piping not otherwise described above	Marley PF501130 P,M,S,G	60	110	$5,440	$5,700	$10,951	$3,900	$10,390
10	TA 1 and 2	P2.11	Compressed air from tie-in at B6 to TA 1&2 and across rail shed	P5.01, P7.01, P0.01	720	449	$25,390	$6,353	$2,150	$5,100	$38,993
11	Filling Area	P2.11	Compressed air from tie-in at B4 B-F, 2-3.5 (Drops as indicated)	P5.01, P7.01, P0.01	1080	340	$22,508	$13,129	$2,150	$4,050	$41,837
12	Mix-Blend, CIP	P2.11	Remaining CA from tie-ins near B4, C5, and D5 (presumed) to Mix-Blend, CIP and drum stations	P5.01, P7.01, P0.01	710	479	$25,896	$9,466	$4,150	$5,500	$45,012
13	Filling Area	P2.11	City water from tie-in at B2.5, to filling area (Drops as indicated)	P5.01, P7.01, P0.01	925	376	$22,759	$20,293	$2,100	$4,350	$49,502
14	Mix-Blend, Almix, CIP	P2.11	Remaining CW from tie-ins at B3A, 4D(2) to Mix-Blend, Almix and CIP	P5.01, P7.01, P0.01	740	655	$35,629	$23,703	$4,100	$7,050	$70,482
15	Mix Blend	P2.11	Install Nano water, complete as indicated.	P5.01, P7.01, P0.01	376	268	$15,756	$57,415	$2,250	$4,200	$79,621
16	First Floor (all)	M2.11	Install all steam and condensate piping as shown	M5.01, M0.01, M7.01	1200	853	$50,746	$41,916	$2,250	$8,600	$103,512
17	First Floor (all)	M2.11	Insulation for above	M5.01, M0.01, M7.01	2300	1120	$81,564	$14,235	$3,227	$1,457	$100,483
18	First Floor (all)	M5.01	Installation of remaining steam / condensate stations, traps, etc.	M0.01, M2.11, M701	NA						$0
				Totals:	11208	9187	$562,070	$306,715	$46,191	$88,118	$987,733

Since we already did some analysis on the state of our job after 15 days have passed, let's use this information to determine what we can pay the contractor for work done through the 15th.

Recall we verified by walkabout the following progress:

SOV line 6 100% complete 88 hours consumed
SOV line 9 32/110 hours = 29% complete
SOV line 1 480/1743 hours = 27.5% complete
SOV line 16, 18 480/853 hours =56.3% complete

SOV line 7 480/2134 hours = 22.4% complete
SOV line 10 100% complete 448 hours consumed
SOV line 11 4/340 hours 1% complete
SOV line 13 100% complete 384 hours consumed
SOV line 14 96/662 hours 14.5% complete

Line No.	Value	% Complete	=	Earned Value
Line 6	$10980	100		$10,980
Line 9	$10390	29		$3,013
Line 1	$153185	27.5		$42,126
Line 16	$103512	56.3		$58,277
Line 7	$214275	22.4		$47,998
Line 10	$38993	100		$38,993
Line 11	$41837	1		$418
Line 13	$49502	100		$49,502
Line 14	$70482	14.5		$10,220
Total:				$261,527

A couple of other metrics are revealed:

261527/987733 =	26.4 % of dollars have been spent
15/55 =	27.3% of our time has been spent
2520/9187=	27.4% of our manhours have been spent
3505/11208=	31.2% of our linear feet of pipe has been installed

In the aggregate, we can claim with mathematical certainty that our job is essentially on track to finish on time and on budget. The line by line SOV curves I challenged you to prepare as an exercise will provide the actual data leading to the same result.

Finally, one more thing we will do before closing this chapter is to go to each of the drawings and highlight the work that has been completed in our first 15 days. That way, we record with certainty, on the record set of drawings where and when each section of work was completed, and as importantly, paid for. Highlighting the work completed on the drawings also focuses our attention on the work remaining for the future. We might also delete or strike through the completed items on the MPL as a tool to keep focus on the remaining effort.

CHAPTER 10

Ideas upon computer algorithms, and how to build a fast track schedule.

The reader is instructed that the discussion on schedules to follow, as a general rule, will proceed from the detailed and specific to the general, building the latter from the former in keeping with the logical construct of building our job from discrete units, blocks, or elements, such as a man-hour. We also develop further the concept of work that can be done in parallel with other tasks or work that must be done in a serial fashion. We also introduce the concept of hard or soft critical path items.

We have until to now focused our discussion on a single hypothetical project with enough tasks and sufficient complexity to illustrate how we draw information therefrom and use it to measure progress, take corrective action if needed, and make payments that accurately reflect the value of the completed work. We included a small rigging scope SOV lines 6&9 and made a claim that this work should be completed before any associated piping work could begin. In fact, if it were possible to install the pipe we would have already done so and say that completion of the rigging is a "critical path" activity, one which must be completed before subsequent activities can follow. We might have been able to run a larger header pipe in the ceiling, but forced to wait to do the drop piping until after the equipment arrived to make a connection. The reader can think in terms of a hard CP activity, and a soft CP activity. In these terms, a hard CP *must* be complete before a successor begins, e.g. delivery before installation. A soft CP contains a certain degree of *convenience*. It might be nice, and more efficient to complete all of the predecessors to a soft activity before going ahead of other work, but a hard CP activity *must* be completed before a another can commence, for example the arrival of welding rods before welding can begin, etc. Workers have to show up to work before any work can start at all. In addition a contractor who is unfamiliar with fast track construction might say that all the hangers must be installed before the pipe can start and all of the pipe must be installed before it can be insulated.

If we have a small scope and a relatively limited number of linear feet, this can be true, efficient and convenient. Work can proceed in a *serial* fashion.

If the job is schedule driven we must attempt to do as much work in *parallel* as possible, i.e. we get the hanger gang out of the gate, and then chase that crew with the piping, and chase the piping with the insulation. In a very simple case with well understood productivity rates our goal might be to land the last hanger on the last day at 10 am followed by the last joint of pipe at 1 pm, followed by the last bit of insulation at 4 pm. Whether we have 100' or 10,000' to install, the concept is the same and just as doable. We know enough about what we are doing with productivity and our labor resources to drive the end date of these 3 sub-scopes to convergence at the end. If you recall back in chapter 5 part 2 we reached a conclusion that the insulation might go five times faster on a per foot basis than the piping. A good assumption then would be that 3 insulators might be able to keep up and stick close behind 15 fitters. A less experienced scheduler might have introduced 3 critical path items on the schedule – (1) hangers complete, start pipe. (2) Pipe complete, start insulation. (3) Insulation complete, start labels. Each of these tasks having hard CP qualities, in the mind of the scheduler, when in fact they are not.

There are plenty of scheduling applications out there, most notably Primavera P6 or latest version and Microsoft Project. It goes without saying that these are useful, but they tend to systematize the scheduling effort, and discourage creative thought. In our example above we have accepted our scheduler's opinion that each of these activities have hard CP qualities. And then, throw in a few more: all the materials must arrive, we must have adequate welding machines, labor, man-lifts, access, etc. And before we know it, our schedule for this simple piece of work consumes 20 lines in P6 or Project, it has lost some richness, and certainly in a meeting with upper management or owners, this kind of detail might be viewed as as overkill. On another hand, a micromanager might focus on an unimportant group of lines and beat us with them if we are behind on anything. In the hands of the Pipefitter Foreman however, we have delivered, on a single sheet of paper, a comprehensive *list* that is expected to be done over the next week. It is perfect information in the hands of the intended audience. An even better job would be to man-load the schedule on a daily basis; showing materials on hand Monday & Tuesday, Hangers working with four guys Monday afternoon and Tuesday, Piping with another four guys Tuesday afternoon (Hangers Complete), eight guys pipe only on Wednesday and Thursday, Pressure testing Thursday afternoon, over time if necessary, and all insulation from start to finish on Friday. The reader can think of this latter a *list and schedule, Level IV,* the former, a *list and schedule Level III.*

Now, let me suggest that schedules with this degree of detail should, as a rule of thumb, fit within the following framework:

1. Show the work of a single craft only
2. Be limited to minimum of 1 day to a maximum of 2 weeks activity when 100% complete
3. Be measurable against a unique SOV line item
4. Probably not exceed 160 man hours (20 guys for two weeks or less)
5. Be issued once a week, looking ahead 2 weeks, and verifying that last week's work got done.
6. The level III or IV schedule should be developed jointly between the CM and subcontractor, as a required administrative requirement after the award has been made.

Again, these six ideas represent a guideline only. They are presented as an extension to the idea where we presented the man-hour or man-day, recorded on the DFR as a fundamental building block or UNIT of work. Now we return to the idea that time is money, and put a price on this schedule $20x40x2x75 = \$120,000$ for Labor and guess \$30,000 in materials. So this schedule becomes another activity for which we can construct an EV curve for measurement and control. If we are working to a two-week period, we can make corrections if we discover a problem in the first two or three days by going to overtime quickly. If we are working to only a single week, there will probably not be enough time to make a correction, but we could probably go on OT quickly and ask for a full Saturday, if necessary, thereby keeping this piece of work under control in the time domain. If we have a budget line item for "unscheduled overtime as necessary" we can draw against it, if not, we will have to swallow an unnecessary financial variance against the SOV line item.

A level II schedule might represent the work performed by one of several subcontractors on a single project. The MPL schedule we have presented above might assume that we are working with a mechanical contractor who can do a limited amount of rigging with his own forces and can supply us with the go-to team if the owner does not want those guys under their scope of supply. In this example of the scheduling level, our structure might require schedules of a similar level of detail for civil/concrete, electrical, sheet metal. Some confusion can be expected at this juncture because a level II schedule from our point of view is in fact a level one schedule in the eyes of our mechanical and other subs. Not everyone gets this all at once. As we demonstrated a guideline or use for the level III and IV schedules, we also have specific properties and audience for the Level II, these are:

1. The level II schedule is a building block of the level one or Master Schedule.
2. The level II schedule should be presented with the bid, by the contractor as a required submittal, in accordance with the schedule of values, as developed the bid documents by the construction manager, project engineer. Or owner's representative.
3. The level II schedule should cover all work to be performed under each subcontractor's contract, and be man-loaded.
4. The level II set of schedules from all of the contractors should be examined to avoid time domain interferences in progression of the overall work. This is one of the duties of the construction manager.
5. The construction manager then integrates all of the level II schedules into a master schedule and issues it to all players for information and reference.

At least one or two iterations should be made by the construction manager in coordination with the other subs, so that the combination of the schedules along with the integration of the equipment delivery schedule, and other major milestones can be well understood by all actors in the development of the level I or master schedule. This step in our process should be used to establish the last of our mutual expectations, which is that when the contractors mobilize they can appreciate what each of the players are up against and shall be expected to work in harmony. The level 1 schedule is the schedule that is posted on the wall of the construction office and or in the owner's conference room and is communicated in the form of a pdf. Nobody gets to change the master schedule unless it is by the construction manager, or by the project scheduler (if we have a budget for one) at the order of the CM. The next thing to consider is whether the architectural and engineering schedules can be considered complete or should be maintained. If those efforts are complete we can drop them. As time marches, the master schedule, by its nature should be a document is diminishing. We do not need to carry a 1000 line schedule when half of the items are completed. In my own experience, it is best to drop lines from the schedule as they are completed. As the work fires up, more and more actors will enter the project domain, and we need not be asked any questions about work that is complete because we failed to get them off the schedule for this week's meeting. Of course we keep all of the proceeding master schedules saved in a common file so we have a record of changes. There are many re-issues and revisions to the master schedule, but the end date shall not be permitted to change unless there has been a *Force Majure* event.

ASIDE

At the Department of Energy at Savannah River I had the experience to work in the area where nuclear waste was being melted and liquefied for permanent storage. Our two week schedule would run to over 1000 lines and 70 pages, (about ½" thick on 8-1/2" x 11" paper) and would cover the work of about 20 managers directing a staff of 200, and 5 schedulers. We gathered once a week on Mondays for an hour-long review meeting, and then the schedule was re-issued every Thursday. We were partly governed by a management culture that wanted everyone to know what everyone else was doing. A much more elegant way of showing this level of information (after years of planning) is the schedule of the Normandy invasion in 1944. Churchill instructed Eisenhower to put it on a single sheet of paper "so the generals could understand it". Presented below is the schedule for Omaha beach. There are similar documents for Gold, Sword, Juno and Utah. The horizontal axis shows the landing zones and the vertical axis starting in the upper left represents time beginning at H-hour minus five minutes. The symbols within each wave represent a force at the company or battalion level.

David Glass

END OF ASIDE

Another attitude among project managers is that there is some kind of floating "baseline" schedule that can drift to the right as new requirements emerge and the end date is extended. This is another matter of project attitude, which in my opinion represents a certain lack of discipline. If there are new requirements that extend the end date, it is much better open a new subproject with its own schedule that does not affect or relax other parts of the job. For example, if there is a change to a mechanical scope that legitimately extends the mechanical schedule, we do not relax the other level II schedules unless those scopes are on overtime and we might realize some savings. In other words, a change in one Level II schedule, which might be a final constraint in the Level I project delivery, does not automatically relax the other level II schedules. Time and time again we witness schedules that have a tendency to swell to the allowable period of performance. It is usually better to put this slack in your pocket until unknowns are eliminated. We do not relax the electrical schedule when we face unknowns in I/O or debug. We do not extend the mechanical period of performance when there is an electrical schedule extension if it remains that we must still complete pressure testing, hydraulic flushing, or other hard CP items in the mechanical contract.

In other literature, and / or computer subroutines, these schedules, which we have described as a level I, II, III, or IV schedules, each showing more detail than the prior, may be automatically parsed out from the level above. The trouble is that the logical structure of the program is one where the information is filtered out from the top down, rather than being built up from the building blocks in the level III or IV schedules. To do this, the automated programs encourage or require more items to be linked than necessary for control of the end date. In the ideal world where we combine the perfect human programmer with the perfect algorithm, a change in one line item automatically trickles down until all slack (or float) is consumed and the end date is *mathematically* pushed out to the right. I appreciate the power of the aforementioned programs, but we must keep in mind that the master schedule will have a diverse audience and a meeting might easily fall off-topic into a discussion about the logical structure of each line instead of whether that particular line has anything to do with what is actually taking place, or if the necessary predecessors have been completed. The tail (or program) begins to wag the dog (management), and we lose sight of the idea that, yes, there are a continuous number of links between the start and finish of a project, but there is more than one way to get from here to there, as we might be diverted with a traffic detour. The detour might be temporary, as in a fender bender (aka a gaper delay in Philadelphia) or one of more lasting nature such as a bridge rehab. There are hard and soft CP items as well as hard and soft predecessors. The schedule lives and breathes from start to finish and is not set in stone. The scheduler and the construction manager work together and in harmony to keep things moving and avoid disruptions. Having gone out of our way to avoid the so-called change contractor, it will take maintenance of the relationship to keep it that way.

There are also such things going on with human behavior such as secret float, secret resources and secret schedules. The scheduling programs also make it easy to insert tasks, trivial or not, and have no way to screen for *GIGO*. If you do not know what GIGO is, you may have been born after 1990. Take a moment and go google it. Modern algorithms have come a long way in the last 35 years, but since the advent of the PC, we still have to deal in units of measure. An hour, minute or second are measures of time; as are millimeters, feet, meters, and miles units of length, and pounds, kilograms, tons, or *gasp!* metric tons units of mass. All subject to the application units of measure and conversion factors. We also must put up with the very pesky unit called the man-hour. As long as these animals exist within lines of code we are still subject to I/O and syntax errors, and as long as there is a human interface between man and machine there will be some big mistakes. The human must set and control the application of the units of measure. If you speaking in kilograms, and your subcontractor is hearing pounds, she might have to run out at the last minute, order a bigger crane and pay for the one that gets used and the one that just gets in the way. Refer back to axiom 12 and verify all the units. We asked for a rigging plan in our list of submittals for a reason. Review it. It is downright irresponsible to ask for it and then just stick it in the file.

Problems with I/O errors at the human interface persist even with cutting edge algorithms. There is an awkward tendency in large corporations to believe in the concept of an "Enterprise Solution"- Upper management, having spent corporate treasure on an expensive software license and training, now must seek out the less expensive technician level schedulers who know the keystrokes specific to the algorithm, rather than those more deeply qualified schedulers with advanced knowledge in the field of operations research and enough school of hard knocks experience to actually know how long it takes to get a piece of work done. In Tracy Kidder's book on *"The soul of a new machine"* (1981), there were these programmers at Digital Equipment Corporation who worried that the introduction of barcode would mean that a cashier in a supermarket would no longer have to be concerned with how many cans came in a six-pack of beer. Isaac Asimov once said "I do not fear computers, I fear the lack of them" a corollary for today might be more apropos: "I do not fear computers, I fear the people using them"

ASIDE

A recent project I was involved in required the manufacture of a $500,000 motor control center. The order was placed with one of those *G*iant *E*lectrical manufacturers, and a change to just one of many of the drives in the MCC resulted somehow in the manufacture and delivery a of a second MCC, a $500,000 GIGO mistake that went unnoticed until the correct one was installed and the second lingered in the warehouse until the project was being closed. We called the vendor and tongue-in-cheek said it was to be subject to storage charges or would be parted out and sold on e-bay. That's what it took to find the problem, and get some humans back on the stick. I'll not accuse a specific company for being responsible, but the enterprise solution of choice at this electrical giant was also known as *S*et *A*side *P*lenty.

END of ASIDE

Dwell for a moment on our prior example having to do with the transcontinental railroad tunnels through the Sierra Nevada, where the drilling of a central shaft enabled excavation on four faces at the same time instead of two, theoretically doubling the amount of *available* work. Imagine our piping job working in the same parallel fashion as previously described, but now with one crew at each end, and two more working from the center towards both ends. Did we just reduce the time required to completion by a factor of four? You betcha. On the downside, we just introduced a handful of new battery limits. Now open your mind a little and consider construction of the interstate highway system and how many discrete segments might have been going on at the same time. In this example, the amount of work that could get done was only constrained by available funding. What are some other constraints worthy of consideration?

Here is a short checklist:

1. We are constrained by our imagination, creativity and learned behavior.
2. We may be constrained by forces of nature.
3. We may be constrained by availability of materials.
4. We may be constrained by availability of equipment.
5. We may be constrained by availability of labor.
6. We may be constrained by availability of funding.

7. We may be constrained by lack of completion of necessary predecessors.
8. We may be constrained by the availability of a mandated inspector/inspection.
9. We may be constrained by a permit delay.
10. We may be constrained by available work space.
11. We may be constrained by activities of other contractors.
12. We may be constrained by availability of front line supervision.

Once we have a solid grasp of our constraints, we proceed to *saturate* the work area with labor. A good example is to observe the beauty of what goes on during an F-1 or NASCAR pit stop, when there is just the right number in the crew. One less and we take more time than necessary to get the job done. One too many and he or she just gets in the way. Of course our considerations are a little different, but the principle is the same: do as much work in parallel as possible with the right number of people. Change the tires, Refuel. Clean the windscreen; give the driver a drink, and go.

We build the schedule line by line, each item saturated with manpower, and then we take this ideal result and pass it through the filter of constraints (above). Does this quantity and quality of labor exist? Do we run the risk of completing work early and have it just wait for inspection? Do we run the risk of going so fast that we run out of materials? Do we have enough men, but not enough in the way of shovels?

And so, an iterative process emerges. We start with the ideal, and pass it through the filter until we bump against a constraint. We check it, and relax it if possible. Getting more shovels is one thing. Expediting a permit or getting the inspector to come early may be beyond our control. If the quantity of shovels is easily remedied and the constraint relaxed, we do another iteration through the filter and another constraint is encountered. Enough shovels but not enough wheelbarrows. We repeat this process until we reach an acceptable or efficient result, and come to the point of diminishing returns for the particular task. For more on this subject, the reader is referred to a body of knowledge known as Operations Research.

As a final illustration, let's imagine the constraints we might encounter in the construction of a simple brick wall. We will begin by assuming the brick oven operated by a vendor has more than enough capacity to serve our needs. The bricks are available in the yard and cool to the touch.

The year is 1928. Hod Carriers bring the bricks to the bricklayers.

1. We may be constrained by availability of credit at the yard.
2. We may be constrained by the capacity of the yard to load our wagons.
3. We may be constrained by the number of wagons available.
4. We may be constrained by the number of horses available to draw the wagons.
5. We may be constrained by the number of teamsters to drive the horses.
6. The distance from the yard to the wall may limit the number and use of the wagon teams.
7. We may be constrained by the number of hod carriers available for work.
8. We may be constrained by the number and length of the ladders available for the hod carriers.
9. We may be constrained by the availability of mixed mortar (Loop steps 1-8, trading bricks for mortar, repeat)
10. We may be constrained by the number of bricklayers available for work,
11. We may be constrained by the number of bricklayers that may be deployed: long wall many, chimney few.
12. We may be constrained by the number of trowels available for use.
13. We may be constrained by the weather.

And now the year is 2016, technology has transformed our process, but:

1. Is the same
2. Is the same, except our wagon has become a diesel stake body truck.
3. Is the same
4. Comes with the wagon
5. Is the same
6. Is the same
7. Our Hod Carrier has become an off-road telescoping forklift, (a LULL, typically) driven by an operating engineer
8. See #7

LULL photo courtesy JLG corporation.

9. Is the same, except the mixer is driven by an electric motor
10. Is The same
11. Is the same
12. Is the same
13. Is the same

And while we might conclude that technology advances have reduced the total amount of labor in lines 1-9 by an order of magnitude, our final constraint is the amount of space available (11). As long as bricks are laid by hand, the amount of time it takes to get the job is the same presently, as it was 90 years ago.

And by now I trust you get the picture. The consumption of time from start to finish is the same. The activities – relative consumption of labor and capital are vastly different.

As a basic point of understanding this chapter, let's return to a few basics of what we need to understand in order to build any schedule. It does not matter whether we wish to create a master schedule for the whole project or a level IV schedule describing a flow of work that should be complete by the end of the day.

BASIC CONCEPTS:

Understand the difference between a hard and soft predecessor, and/or a hard and soft milestone. It takes experience and some good examples to know the difference; which ones we need to pay attention to and which ones can be taken for granted. I gave you an example of workers showing up for work, and we can take that for granted, less a degree of absenteeism or a big disruption in the labor market. Another was the availability of welding rod before welding can get done. You cannot drive on wet concrete is another example, etc.

So, here are the basic steps to building a fast track schedule without the application of overtime:

1. Understand what can be worked in parallel, vs. what needs to happen in a serial fashion.
2. Understand the point at which a work area becomes saturated with labor and the point of diminishing returns with the application of more of the same.
3. Understand the very moment a new scope of work is available and ready for work. Once we have saturated a work area, we must strive to maintain it in that condition. Keep the daily manpower as constant as possible for as long as possible, while recognizing every job will start slowly, reach a peak level and then decrease as more tasks are completed and fewer can be introduced.
4. To the extent of your natural ability, think of your project not so much as a series of critically linked tasks from beginning to end, but as a series of sub projects or work breakdowns that have discreet starts and finishes within the total job. Cross things off when they are done.

5. Forget what you might have learned about shifting baselines, instead make a new subproject to stand alone if you need to. If you keep a file of the level 1 schedule as they are revised you can always do a forensic study if needed.

6. The Master Schedule is built on the Level II schedules. Go ahead and create a master schedule for axiom 2 (planning) activities up to the point where preconstruction ends, then start another where construction begins, but keeping in mind that planning is an ongoing effort that does not end until the job is done.

7. The key milestones of our master schedule(s) START and FINISH should not be changed except by *force majure*. We can revise or tinker with the guts all we want as long as we avoid disruptions. Think of it as a matter of management attitude.

8. The Level III and Level IV schedules are your *tools* for getting the Level II work completed on time. These schedules should expire in relatively short periods of time and form the basis for your financial control as the SOV lines are completed.

Now, having maximized the work in parallel, and saturating the work area with adequate craft, the only thing remaining thing we can do to shorten the schedule at this point is to put our labor force on overtime. Continuing with our schedule at hand, we will demonstrate how to do this in the next chapter.

CHAPTER 11

The mysteries of overtime are revealed.

There are plenty of studies across the board demonstrating that overtime, as we shall define it as work beyond 8 hours a day or 40 hours a week is generally a bad idea. And that there are diminishing returns as the percentage of work done on a second shift (within the same constraints) increase as a percentage of the total[1]. Keeping in mind that our discussion is confined to construction rather than manufacturing, there are only a few situations where overtime is more often than not justified; mostly situations where well-known, highly profitable manufacturing operations are shut down for maintenance and need to be back on line as quickly as possible regardless of cost. Examples would be a power station boiler rebuild, some mining operations, and refinery turnarounds. These are always schedule driven jobs where the cost of downtime is always in excess of the cost of labor to keep things running regardless of labor cost or inefficiency.

We return to our Net Present Value analysis as the guide to our decision. We return to the idea of *Schumpeterian* products which are characterized by high early prices willingly paid up by early adopters, followed by rapidly falling prices, then obsolescence or creative destruction. Today, we see most commonly these things as fax machines (obsolete), VCR's (obsolete), CD's (almost obsolete), tablets (rapidly falling prices), and the latest smart phone (expensive and sold to the early adopters). And batteries which are pushed out of the market by successive generations of the same product – continuing improvements in the watts/mass deliverability coupled with falling prices due to economies of scale. Imagine a decline curve on revenues where we might get half of them in the first two months out of a product having a life of two years. Winner takes all; second place is the first loser. A race to market is thrust upon us.

Now, having set the stage, and having made a previous presentation on the subject of net present value analysis, it remains the owner's prerogative to set the schedule and determine the requirements for overtime. The compressed, fast track case is different only in that the project and resulting product lifecycles are very short. Previously, the project might take a year or two and the product lifecycle be measured in decades. Now our project will take less than a year and the product lifecycle might be done in 18 months or three years. In light of existing literature and our own experience we will now explore what we can get my compressing the schedule, by first going to 50 hours a week, then to 60 hours a week, then the addition of a second shift, the latter of each being less efficient than the predecessor. We are already working on an efficient parallel schedule as suggested in the prior chapter. The math is relatively straightforward. We do not change the scope, just shorten the time allowed to get the job done, and consider the marginal cost of premium time, a shift premium and some assumptions about different rates of productivity.

Going back to our man-loaded schedule we just run a few cases, and for simplicity assume we are prepared to do each case on time and materials:

Recall our rate sheet introduced in chapter 6 had a bottom line of

Straight Time = $76.98 per hour
Time and a half = $91.21
Double time = $112.44
Man-hours Required to Complete the work 10056 (a constant)
Average crew size 22.85 men (a constant)

The cost of labor on straight time, is 10056 x $76.98 = $774,110.90

The blended rate for a 50 hour week is (40 x $76.80 + 10 x $91.21)/50 = $79.68
The cost of Labor based on a 50 hour week is $79.68 x 10056 = $801,282.20

The blended rate for a 60 hour week is (40 x $76.80 + 20 x $91.21)/60 = $81.60

A the cost of labor for a 60 hour week is $81.60 x 10056 = $820,603

Now, recall we made the assertion that we have our work areas fully saturated, we have the means to keep it that way throughout the job, and our base case has an average crew of 22.85 men in the time domain… our "burn rates" for labor shake out as follows

On 40: 10056 man hours / 40 hours/ week x 22.85) = 11 weeks 77 days
On 50: 10056 man hours / 50 hours / week x 22.85)   = 8.8 weeks 61.6 days
On 60: 10056 man hours / 60 hours / week x 22.85) = 7.33 weeks 51.3 days

To look at this on a calendar assume we start on a Monday, January 1:

On 40 complete COB 9 March
On 50 complete 2 March after lunch
On 60 complete COB 20 February

Time is money, and we have just demonstrated how much we can buy with it.

We shall turn this information over to the production and finance people and they will look at their expected weekly gross profit from the new work and compare it to the marginal cost and tell us how to proceed. Since the time frame is so short we only need to do a modified NPV analysis, because the compounded cost of capital over the period is small. *If the daily gross profit exceeds the daily cost of acceleration, then they should want to go fast.*

Having done the math, we turn our heads to some additional factors for which the math is tricky, and not perfectly solvable in general, but we can say two things in general and more or less for certain:

As the hours per week increase beyond 40, productivity will diminish at an increasing rate as the work force begins to tire. These rates will differ between the crafts, and the task they are working. Absenteeism will fluctuate because some people only want to work 40, some want to work all they can get.

We have before us a problem of diminishing returns on labor and no demonstrable way to solve the problem, except by direct observation of when the derivative curve shifts to the negative, for our particular case. We can at least provide a list of things to watch out for.

Hot or cold outside work is by nature less productive than nice and cool inside work on a flat surface. It rarely makes sense to try to accelerate work when the weather is uncooperative.

As the length of the overtime period gets longer, the productivity will decrease and the absenteeism will increase at an increasing rate. Workers that need the work for financial reasons will afford some time off the longer they are on the job. Some will just want to do something else for a change. Depending on the amount of training, testing, safety certifications, and other corporate or government requirements, the new worker might not be "productive" or even available for productive work for a number of weeks. Having spent some time working a CM contract at DOE Savannah River, a federal project with the highest requirements for on-boarding craft workers I ever witnessed, we used a rule of thumb based on statistical evidence that a typical craft worker who quit his job cost about 160 hours in non-productive replacement costs.

And now, we look at the idea of going on a second shift, first the problems. We will introduce some basic ideas, which the reader is asked to take at face value.

In many union agreements, there is a contract rider that requires a shift premium commonly 10-15% that goes to the worker.

There is empirical evidence that second shift work is less productive than single shift work. We face the issue of the same scope of work having two front-line supervisors. As work is handed off from one shift to the next, time is required to communicate the progress and exact status of the prior shift. If the shift is 8 hours, the foremen might need to work 10,- a one hour overlap on each end to communicate what happened before. Among the craft, there is the problem of picking up where the prior shift left off. The tools aren't exactly where they left them unless the job is big enough to have a tool-keeper and brass system in place. This is disruptive for all of the journeymen but for a shorter period of time. In other words, *a day-to-day destruction of learning curve effects is evident.* Serious issues might cause a stoppage in one area due to the unavailability of higher-level supervision, resulting in idle time and workarounds. And finally, we will leave our reader with a study[1] that demonstrated the productivity on a second shift decreased as the ratio of second shift hours to first shift hours increased:

Productivity Loss = 0.22052 + 0.07152 ln (% shift work expressed as a decimal).

Now, the resulting curve would be based on the data within the study and the nature of the work performed, but let's go ahead and examine a case where our sample job encounters a scope change requiring 20% more work without change in space or extension of schedule. All else is maxed out and under these conditions we are forced into a second shift.

We use 10056 as our total hours, 20% = 2011 hours

Inserting 20% into the equation our productivity loss is:

DP = .22052 + 0.07152 ln (.2) =.1054

Now, we take the productivity loss across the hours which are overlapped, - in our case 2011 hours on the first shift and 2011 on the second shift and we have:

Total Loss = (2011+2011) x .1054 = 424 hours.

And as we are already working this job on 6-10's our effective labor rate is $81.60 so value of the work that has evaporated is $34,598. Think of this as a real penalty cost for keeping the schedule. If you are faced with this math as a real added cost, stick this penalty into your NPV analysis. Might just be better to relax the schedule, but probably not. As owner of the next-gen gadget, how well do you know your competition?

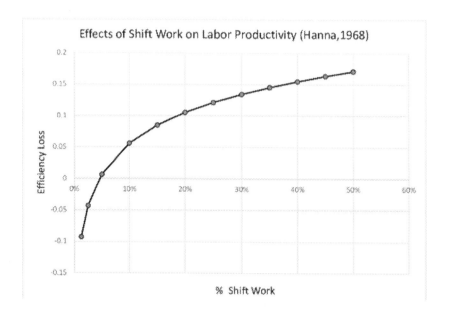

Chart, credit Hanna 1968

Now, let's inspect this curve a little further. At around five percent overlap the sign of our function changes from negative to positive, just as we might expect given that the natural logarithm function changes signs at 1.0. ln(x<1) is negative and ln(x>1) is positive. In fact the function is solved for 0 at about 4.6% overlap. Not a big number, but if we are truly in this race against time and need every hour we can get; the math speaks for itself, *res ipsa loquitor.*

And by no means are we suggesting that shift work is some kind of demon. All we are suggesting is that in this case, for the 2011 extra hours requested, there is a hidden cost of $34,598. Seventeen Dollars and Twenty Cents an Hour, for this particular job, at this particular point in time, and this much overlap required.

Now, to understand when we should go on OT, picture areas of work where other more, *disruptive* conditions exist and there is no overlap. Picture ductwork that takes up lots of floor space and gets hung in the ceiling relatively quickly. The gang of Tinknockers consume aisleways and tend to block traffic. It very well could make sense to put all of this craft on a second shift as long as they clean up after themselves, and leave the site the way they found it on a daily basis. It could also make sense for the insulators, who take up space and generate a lot of scrap. And then, there are the painters, who have to cover the work of others to do their job. In these cases, have a *gasp! a plan, a budget!* for working a second shift.

We close with 3 more examples where shift work is desired, if not required.

Imagine we have to hydro-test a length of pipe after it has been installed but before insulation. This work is best done after-hours because of the possibility of leakage and the uncertain amount of time involved in corrections. We plan for two hours, but wind up needing eight. In some summer locales, the lost productivity just due to the heat associated with outside work and the danger of heat related illness could dictate second shift work just because it is generally cooler at night. Finally, some work will require a utilities disruption, such as a transformer change, or a punchlist of things that can only be done during an operating plant shutdown.

And as we close this chapter, let me suggest that we revisit the bridge analogy (Ch5, p2) on the cost of getting a late or slow start, and how the cost of correcting the problem grows faster the longer the lag. From everyone's (Owner, Construction Manager, Contractors) perspective it might make sense to hit the job running: at some point after all of the craft is lined out with work and the level IV schedules are on a roll, the EV curves forecast clear sailing, - go ahead and work a few weeks on OT. Get ahead of the game and take a stroll to home plate under budget instead of crashing into the catcher in the bottom of the ninth, down by a run.

1. Journal of Construction Engineering and Management, March 2008. Awad S. Hanna, Impact of Shift Work on Labor Productivity for Labor Intensive Contractor.)

CHAPTER 12

A Brief Note on Budgets.

We have noted that time is money on more than one occasion and we followed the jobs progress from start to finish accumulating man-hours and costs. The budget itself is another kind of animal. We even went as far as to state early on that one should avoid a management position in a project unless he or she has had some input into the development of the budget. In the classic sense there are 5 main categories to any budget: Labor, Materials, Equipment, Subcontractors, and Other. We have discussed at some length the nature and cost of labor, and the difference between construction and capital or utilities equipment, and the hiring and behavior of subcontractors.

We demonstrated that time, being the same as money can be purchased up to a certain limit by applying overtime to particular situations where work areas are saturated with labor and as much work is going on in parallel as possible. We also made note that early on in the project, while the scope is still being developed, we may see an increase in the capacity requirements of our utilities. We have planned to the best of our abilities, and have developed a budget and schedule where we expect the work to be complete on the day our last project dollar is spent.

We must; therefore, refuse any pressure from the customer or upper management to bring the job in *ahead of schedule and below budget.* Never have there been two ideas in such perfect polar opposition to the natural forces regarding time and money. There can be one and only one budget and schedule for the job with all parties in agreement.

Yes. After all of that planning and preconstruction effort, having crossed all of the T's and dotting all of the I's, someone in a position of power upstairs will ask for a compressed schedule at a lower cost, without an inkling about axioms 1 and 2. They ask, as though there are these little construction fairies with hardhats and magic wands in their tool belts working alongside the journeymen for free. If you agree to such a request, you merely blend into the madness and start the job from a position of weakness.

It is much more likely that somewhere along the way, something will be delayed or cost more than expected. Most often, this will come from acts of nature or unexpected underground conditions. During the construction of the Panama Canal, soil conditions around the town of Culebra were such that the excavation did not stop caving out of control until a 2% angle of repose was reached. The amount of excavation required was several orders of magnitude higher than expected and led to the bankruptcy of the first effort. In almost any other civil work, contractors run the risk of rock excavation or unforeseen conditions.

A typical day at Culebra in 1913

The best that can be done in this case is to set an expectation as to the possible cost of rock excavation. In other situations, the soil may not bear the intended load and the foundation design may require changes. We have all of the tools we need to run a NPV against percent rock, and NPV vs. the cost of some test borings in order to disclose the estimated percentage. At this point, I have led the horse to water, and you have the right analytical tools to set reasonable expectations at to cost scope and schedule.

What I am trying to drive home is the point that the budget is our *comprehensive spending plan* for the job. Type 1 errors will occur, and we may need to draw funds from another line item in the budget. We also make a budget for a line we shall call *contingency*. This budget should be set by the project estimator, within his or her view of risk, not the salesman or manager. We also consider a budget for what we shall call UMI or Unidentified Minor Items. Think of the contingency budget as funds to solve unforeseen problems (the risk) associated with the known scope, and the UMI budget as funds to make the job more pleasant, especially if you are working in a remote area or harsh environment. Few things make a job more pleasant than a bar-b-que grill and a hot lunch with the superintendents and foremen every other Friday. Carefully consider a travel budget for vendor visits and inspections, make sure there is a budget for the maintenance of general conditions, temporary roadways, stairs, platforms, temporary power and lighting. Temporary fences, barricades, security, environmental protection and controls – dust control on one hand and washouts on the other. The US Army corps of engineers keeps detailed records of weather conditions for places across the US, and you can write into your specification the number of expected days to allow for inclement weather month by month. Go ahead and put that in the bid package and you can avoid the contractor who uses the weather as an excuse for not getting things done on time.

Another thing we can control for to some degree is volatility in the commodities markets. Steel, copper, fuel, and aluminum prices can vary greatly in the short term. Most contractors will put a line in their bid to wit: "due to the volatile nature of steel prices, our bid for this item is good for 7 days." Then they neglect or forget to document the expected cost per pound or tie the price to a mutually agreeable price index as published in the Wall Street Journal or some other trade publication. If you want to get precise, estimate how much material is at short term market risk, tie the quantity to an index and price, and then make an adjustment for future changes at cost. We mentioned early on that our contractor is not serving as the project bank. She is not serving as our hedge fund manager either.

And we will close this brief chapter with a remark on payments. Earlier we discussed money as a tool, means, or grease, to lubricate and move the work forward. As the CM or PM or owners representative, you, and you alone should control the purse. You have the EV curves at your disposal, which will tell you on a daily basis exactly what the contractor is due. If you agree with the contractor to pay net 7 or 10, you must make it clear with the guy upstairs writing the checks that the terms of payment are just as much a part of the mutual expectations as is the scope of work. Time and again we witness corporate paymasters bring work to a near standstill because they have every incentive to make money on short-term investments; paying the contractor on time is never very high on the list. Late payments or breach of financial promise, even when made verbal or in passing will cut through your credibility like a hot razor. You examine each invoice carefully, reject it if necessary for corrections and only send fully approved paper upstairs for payment. You rocked the boat when you came on the job, and now, with your non-routine request to transfer money from the controller's office to this outside actor is going to cause some issues until you have gained a level of comfort with the paymasters. If you are really on top of the EV curves you will make these new friends by forking over a monthly projection of the outbound cash flows so no one is surprised out of ignorance or lack of interest or attention when the invoices start coming in. You may very well be dealing with a finance or accounting department that never met you and has no idea you are in a temporary office downstairs, Their management has asked them for a statement of project cash flows, and they might be able to guess, but if you run the information upstairs without being asked for it, you will make a new friend because you have placed he or she in a position of taking credit for your complete and accurate forecast.

CHAPTER 13

So you want to be a construction manager, or, Oh! the places you'll go.

The end of the job can be one of the most satisfying, or dissatisfying, most difficult and always the strangest periods of any project. Imagine if you will, what has just happened. A group of people sitting in a boardroom came up with a budget for the next-gen gadget. They decided they would need some kind of project manager or construction manager. Low and behold, there you were on the internet, on in the database of a job-shop. You were hired as a temporary tool to get a job done. In the beginning you may be invisible because all you do in the form of planning and vetting the actors is off site or down in the basement. The scope of work you are doing is known only to a few people near the top of the company. Then some accountant is faced with your first weekly invoice for $5000 (which is more than almost anyone is making, but also, unbeknownst to said accountant also covers your temporary cost of living, miscellaneous expenses and travel). She, having no idea what you have committed to decides to meet this new *outsider, maybe even a yankee* – or in Kentucky, a *hillbilly* or *flatlander*. Even though the plant manager signed the invoice, she makes it her business to know where this kind of money is going and delivers the first check by hand and says, her voice dripping with a mix of envy, anger and jealousy "*here's your little check*". And there you are, still alone in the cellar with desk, laptop, and cellphone. You smile and say thanks. On your developing phone list next to her name you make mental note: "probably doesn't like me."

If the job is not that big, you do everything by yourself. Those on the board with the money have many other duties and only want the big picture once in a while or once each quarter. That's why you keep those EV curves up to date. In the meantime, you toil in obscurity and people only come to see you or call on the phone when they need something. Sooner or later management upstairs notices you are working only because the security cameras indicate you are the first one in each morning and the last one out each night. Sometimes your car is there on Saturdays when you catch up on paperwork. One day the accountant brings your weekly pay and asks if you might not like to have direct deposit since you live 2000 miles away. Her countenance warms up a little and she asks about the picture of your children you keep on your desk.

In the next weeks and months we build the head of steam and place our trust in the vendors and subcontractors. In time 500 Journeymen, and a few Journeywomen come on the job, some from the local union hall, some travelers. You try to win them over with a clean place to work and good sanitation, close proximity to potable water, and maybe even microwave to have a hot lunch. You spent $15,000 on the general condition called parking so they don't have to walk through ice cold mud to get to work and you spend a little less shoveling off the road when the neighbors complain. Some of the workers really appreciate your efforts, some just want to get paid and go the bar on Friday and will drag up or come in late and hungover on Monday. There is a husband and wife that come together. Your go-to team turns out to be a father and son. With so many people on site we have a true cross-section of humanity: there are also addicts, miscreants, thieves, pilferers, evangelists, liars, cheats, pugilists, dreamers, poets, troublemakers, and fugitives. One gets his guitar out and plays for everyone at lunch. Another sings in the church choir. Still another just gets his job done and doesn't say a single word or make eye contact with anybody the whole time he is there. You are certain that more than a few are packing heat in their coveralls. One welder emerges as the most talented one on the job, another gets run off because his work can't pass an x-ray examination. Most all are just trying to do the best they can. While you have done your best to vet the subs at the management and superintendent level, the work force is another beast more or less thrust upon you for a limited amount of time, then everyone moves on. You do your best to have fun in the meantime, hope for discipline, and that no one gets or arrested, assaulted, falsely accused, or hurt on your watch.

There is one more type of person who comes on the job at the end. These are the vendor's equipment supervisors and/or start-up supervisors. On a big job, half of them may have English as a second language. Some fine equipment will require supervision throughout the installation and you get to know them a little, some may just come for a few days for checkout and start-up. While you are responsible for all of the activities on the site these people will have razor focused tunnel vision on their own piece of work. In accordance with axiom 2, we have them on a schedule, know what, in the form of completed or temporary utilities they need for their work. If you have not provided at least a desk and chair, or better a temporary visitor's office, you will meet them assuming your desk is a public space, sitting in your chair, using your copy machine, and asking what number to dial on your phone for an outside line. If you were not part of this vendor's procurement, you probably don't have an idea of how much time, or how many trips came with their proposal, or what they are supposed to accomplish on-site. There is also a good chance that their gear or kit shipped by fed-ex to the office address and did not fall into the receiving plan you worked so hard on to put into place, so as first order of business they will immediately need to borrow a forklift, or just get on one without permission to drive around front and get their stuff. They can come on like a bull in a china shop or a prima ballerina demanding your attention. Still, they don't report to you, they report to their own supervisor in Germany or Sweden and have to be in Caracas on Thursday. Some are even, *gasp! independent contractors.* Then they tell you they expected the services of a couple of journeymen for a week to do the I/O checks. If they can't start immediately they have to go somewhere else and can't come back for a month. After all that has been done, you may find yourself hostage to their mercy and they know it. They show you their quotation; there it is, accepted by the owner, but there is no other proof of a budget for the journeymen. We write what we hope is our last Field Order directing the electrical contractor to provide two guys to work at the supervisor's direction for 80 hours. As owner, you cannot turn your nose up at the line item in the equipment proposal called field support. While these supervisors are expensive, and have a monopoly on the service, keep in mind start-up is your final milestone on the path to those early adopters with the big wallets.

For complicated or specialized machinery, we might specify a demonstration run-off prior to shipment, and build slack in the equipment delivery schedule to identify and correct problems. If not, we run some risk of having the machine ship along with some last minute issues.

The point we make here is that until the job is done, we may still uncover a gap in the overall scope of work at the last minute. When we have crossed 5,270 feet of our mile long bridge in 59 seconds, we find that there is spike strip near the end instead of a finish line. When we began the planning process, our attention was focused on problems that were clear and present. Now, returning to our model of forecasting the path of a hurricane, we have done the best we can to locate the eye, but still at the end we might wind up somewhere else within this cone of uncertainty at the very last minute. As foreshadowed in chapter 1, some type one, or even type two errors of omission may have been unwittingly committed before we came on the job. Over the weeks and months that followed we have become identified as the one most responsible for the whole job, and we can get stuck with them. It's called the hunt for the guilty. *Oh my!*

We also have to manage another beast called the punchlist. Having done our best to set the mutual expectations throughout the job, we still run the risk of having new line managers or operators coming on board as a stakeholder in the new future production who had nothing to do with the effort until the end approached. They may try to add items to the punchlist at the last moment. Some good ideas, possibly, but are outside the present contractual requirements of budget and schedule. From the owner's point of view, it gets called punchlist, from our contractor's perspective, in part, a wish list. Things we call "nice to haves". The trouble is that as we near completion, the vendor supervisors present us with a new set of requirements, the customer hits us with a punchlist half of which is beyond our scope of work, and at the same time all of our subs want to get the hell out of Dodge. The transition between construction and start-up can be a difficult period of the project to manage. The owner demands things beyond scope and holds the final payment. If we have had open and honest communications from the beginning, in accordance with axiom 2, the time spent planning and setting the mutual expectations has resulted in a shorter punchlist and greater clarity in which party should be financially responsible for each item.

Beyond the contingency or UMI budget, we might set aside some money for the punchlist. Even if the scope is disputed we keep an option to pay for the last minor item and call it goodwill or a small amount of grease.

And then there is a potential for either equipment that was not properly specified during design, or does not meet a properly prepared performance specification. A robot is not strong enough, or can't reach far enough to do his job, something keeps going into an overload condition, or something else just gets busted. If we have a big job on our hands, there may be another list called a lubrication schedule. Sometimes the oil is shipped in the gearbox, sometimes with the gearbox, sometimes sold separately. Before things start in motion the CM or his project engineer has verified everything requiring lubrication has got it.

COROLLARY TO AXIOM 2

AS WE APPROACH THE END, THE BEST LAID PLANS CANNOT PREDICT EVERYTHING. AS COMPLETION LOOMS, WE WAIT AND HOPE.

Finally, the day comes when things go on-line. The new workers are at their stations; all of the journeymen are gone. The new $500 gadgets go in the box every 15 seconds. You walk about the plant with a sense of pride, having tamed the tiger. The new workers look at you with no idea of whom you are or what you are supposed to be doing. Someone even reports you to the new security manager for not being within the yellow paint surrounding a work station just like everyone else. You are his first bust. You are temporarily detained, then you go outside to find the trailer getting hauled away. The demand curve for your services evaporates and you disappear, just like that, into the next undiscovered country.

LAST OF THE ASIDES (On my travels with Charlie)

The idea for this book began years ago as I grew up in the long and narrow Bald Eagle valley of Central Pennsylvania. A land that was connected to the rest of the world at first by Pioneers and Pathfinders, then a canal, and later a railroad built by the capital of Andrew Carnegie and Robert Pitcarin; not to mention the labor of the Irish and other immigrant labor. The geological barrier, a continuous fold in a thousand feet of solid rock separated the East from the North at Williamsport, where a glacier came through, and cut the path of the Susquehanna. The West, isolated from the South, except for the Cumberland Gap at the confluence of the Tennessee, Virginia, and Kentucky countries which was cut by wind and water over many more other millions of years. The highway near my home was finally cut through the rest of the Allegheny Front to the 4 lane US standard all the way from the old canal take-out to Pittsburgh in 2013. Some company towns along the way were lost for good, but one can still see the much older ruins of the road built through a places named called Portage, and Cresson Summit, where the canal boats were taken out of the water and lugged over the mountain by men using mules and rope, and rails of Iron, but not yet Steel. There are also the Horseshoe Curve and Gallitzin Tunnels dug by hand 160 years ago and still used by Amtrak trains #42 and #43 daily.

In the Northern reaches of the valley, there is a gliderport at a place named Julian, Pennsylvania, where the weather and geology come together to from perfect conditions for soaring. In 1981 Tom Knauff got in a sailplane and flew all the way along the ridge to Knoxville Tennessee and back, at the time a world distance record.

It was along these paths that I escaped the natural confines of the Tuscorora Quartzite, Bald Eagle Sandstone and Gatesburg Dolomite, and found my wife in another town where better men like my father-in-law made steel; a place called Weirton West Virginia. A place where good jobs once gave the town the best per-capita income in the United States.

With the advantage of jet engines, still built by hand, I crossed the Atlantic 10 times and the Pacific 20. I have seen bits of more than 20 countries. At the age of 24, I found myself standing in line at the American embassy in Abu-Dhabi, needing more pages for my passport. In 1985, I saw a building foundation in Islamabad dug by pick-axe and earth moved by mules; the only reason being the labor of men was cheaper than a hydraulic excavator.

Then there was an FAA job which required travel to a different city every week for three years, helping to rebuild the National Air Space we all take for granted before and after 9/11/2001. Places like Pensacola, Gulfport, New Orleans, Barstow, Saginaw, Sioux Falls, Reading, Pueblo, and Maui. There was also Monterrey, where I woke to the song of sea lions, and read Steinbeck. You could see the Salinas Valley change color 500 acres a day by whatever the men and women, bent over their backs were planting or harvesting. There was also a memorable sight of a bale of cotton the size of an intermodal shipping container in the middle of a field after 100's of acres were harvested by machinery overnight.

Somewhere along the road, my bones grew sore and I came to need help loading the suitcase. Soon I will need help just cutting the grass on my own 10 acre piece of ground. I guess arthritis is the unkindest cut of all, and not fair to anyone. My friend and former professor, Dr. Carl Sherman told me "Dave, it's not the age, it's the mileage."

My favorite memory comes from Charlie Myers, that man who spoke of those pesky *waars*, or *warrs,* (however you want to spell them), sticking out of the solenoid valve and the magnet on a stick and the *Arn* bolt, just out of arms reach in the extruder. After the completion of a turnaround at a sprawling yet ephemeral Mitsubishi Plant in Maysville Kentucky, we had our gear loaded and the plant was back on-line producing 50 ignition coils a minute for all their customers; the wheels of international commerce running smoothly once again.

Men who live a life on the road and work in construction do not call others "friend" lightly. Like Cicero on Laelius or Scipio, we more or less come to count them on one hand, having busted a knuckle or two together along the way. My friend Charlie, the aforementioned master millwright and I were about to part ways for the third or fourth time after doing a job together. As a traveler, you almost never expect to work with the same people more than once. There was a certain *Joie de vivre* and look in his eye. I had of grown fond of, him and hope he felt the same way. We shared a kind of self-deprecating Appalachian humor. He was headed back to Hopkinsville in Western Kentucky, and I home to Apollo in Western PA. I told Charlie I was thinking about taking a drive through the coalfields of Eastern Kentucky and whether he had any idea about things to see.

He said, "Now Dave, I may be from Kentucky, but more precisely, I'm a Flatlander". "Them parts you're talking about… well, they's mostly Hillbillies yonder. Most all I ever heard of em is if you ain't kin, you best keep moving".

CHAPTER 14

Epilogue: Breaking the Circle of Doom.

Years ago Dante wrote his famous Divine Comedy starting with the Inferno, followed by the somewhat lesser known Purgatorio and Paradiso. Graphically, the Inferno has been drawn in many forms, one of my favorites presented here:

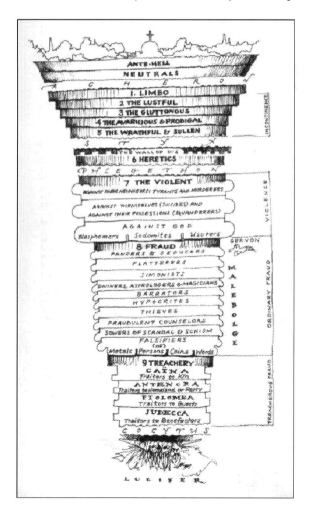

Illustration courtesy of Barry Moser, used with permission.

According to Dante, the only two who that made it all the way to the bottom were Marcus Junius Brutus (the younger) and Judas Iscariot.

If you believe your present writer, I can tell you there are more than a few others, and it goes without saying that if you spend enough time on the road, you will meet each and every one of these other kinds of people.

***

Now, seriously,

We began this book with the tongue in cheek project flow model many have been acquainted with in one form or another during their career:

Let's just go ahead and call this the circle of doom, or another view of the first few rings of a modern-day hell. As this book has progressed we discussed a real variety of means and tools by which we avoid getting sucked into this vortex. First, we avoid the project unless we have had an up-front involvement in the budgeting and scheduling and are comfortable that the project will be successful.

I have attempted to break this circle by a taking a pragmatic look at some old tools in a new light. For years, contractors and owners have been arguing over percentages of completion; ignoring a structured schedule of values for the work to be done; failing to specify, completely the scope of work down to the location and type of every battery limit; and complaining and initiating claims for inefficiencies and delays while working with lower levels of manpower than might be reasonably expected, while not even talking over a properly man-loaded schedule.

We drove home the importance of on-going planning, arm's length subcontractor and vendor selection, the importance of drawing issue and control, tagging and cross-referencing devices, instrumentation, and the drawings themselves. We organized our project area in to spaces for each contractor, a control area for inbound materials and equipment. We spent a fair amount of time on establishing mutual expectations, and controlling scope changes by way of field orders and subsequent change orders. We discussed a variety of contracting schemes: Lump Sum, Cost Plus, Time and Materials, and some hybrids. We presented a few ways to co-opt the disturbances to the daily plant routine and make new friends along the way. We discussed the scheduling concepts and the usefulness of the different levels of scheduling. We discussed the role of the engineer as one who solves problems by breaking things down to their component parts and then re-assembling them to a completed functioning whole. We de-mystified the seemingly complex aspects of the engineering effort, and did the same with business and financial aspects of the job by parsing out a structured schedule of values; call them subprojects if you will, but we have effectively erected business firewalls, or by way of another analogy we constructed a compartmentalized ship separated by a number of bulkheads. A fire or torpedo in one compartment should not result in a total failure of our project. From time to time, we summarized our ideas in the form of an axiom, as I see them; presented herein for what they might be of use to my readers. As a final suggestion, let's revisit Hermagoras and post on the wall of the project office:

Quis, quid, quando, ubi, cur, quem ad modum, quibus adminiculis.[9][10]
(Who, what, when, where, why, in what way, by what means)

And add "with how much," *quanto.*

And at long last, returning to our concept of being able to derive anything from the 3 fundamental units of mass, length and time, allow me to add just one more, constructed from the sub-particles above: that elusive and often ephemeral element we call the truth or:

VERITAS

We ask ourselves these questions about any aspect, actor, or issue on the job, and if the answer is clear, timely, or better yet self-evident, and if our EV curves are pointing in the right direction, then our chain of doom is broken, and our project success is all but *Axiomatic.*

D.E. Glass
April 2016

A Glossary To The Axioms

AXIOM 1

NEVER AGREE TO A SCHEDULE OR BUDGET UNLESS YOU
PLAYED AN INTEGRAL PART IN BUILDING BOTH, OR HAVE
HAD AMPLE TIME TO VERIFY THEY ARE REALISTIC.

Think of it like buying a new house without knowing the price, floorplan or delivery date.

AXIOM 2

IF THE JOB IS TO BE SUCESSFUL, YOU WILL HAVE SPENT
ABOUT 65% OF YOUR EFFORT PLANNING AND LEARNING.

And the rest is measurement and administration. The planning is on a strategic level at the beginning and shifts to the tactical, a plan for the week and plan for the day.

AXIOM 3

ALARMS AND INTERLOCK DEVICES ARE THERE FOR
YOUR OWN PROTECTION AND THOSE YOU LOVE.
YOU CAN HIT THE SNOOZE BUTTON ALL YOU WANT,
BUT NEVER, EVER, IGNORE DISABLE OR DEFEAT
ANY OTHER ALARM OR INTERLOCK DEVICE.

AXIOM 4

WHATEVER THE SOURCE, EACH DEVICE SHALL HAVE
A UNIQUE NUMBER STAMPED ON THE DEVICE OR A
LEGIBLE TAG SUITABLY AFFIXED. THE DEVICE NUMBER
SHALL BE CROSS REFERENCED ON ALL DRAWINGS,
EQUIPMENT NUMBERS, PLANT AREAS, PURCHASE
ORDERS, VENDOR DOCUMENTS AND PLC PROGRAMS.

Cross-referencing is one of the keys to avoid confusion, keeping complex systems organized, and recovering when things are disrupted by the errors of others.

AXIOM 5

AS A *MANAGER*, ALL YOU NEED TO KNOW ABOUT INSTRUMENTATION AND CONTROL DEVICES IS THAT THEY ARE CONSTANTLY IN PERFECT 2 WAY COMMUNICATION, SPEAKING A COMMON LANGUAGE, THAT IS PERFECTLY UNDERSTOOD BY THE PLC WHICH READS AND SENDS ALL THE MAIL EVERY OTHER MILLISECOND.

This basic introduction to instrumentation was to illustrate how much data can be collected, transmitted and controlled in a simple system, and to demonstrate that the interactions between systems can grow geometrically as opposed to serially.

AXIOM 6

DON'T EXPECT HONESTY AND SALES TO COME FROM THE SAME PERSON

Buying a multimillion dollar piece of industrial equipment or contract for services are like buying a used car. They are experience goods and you will never know for sure about the result until the job is done and the salesman is long gone.

AXIOM 7

IN THE RFQ FOR THE PURCHASED PARTS YOU SHALL SPECIFY TO THE VENDOR THAT HE OR SHE SHALL TAG EACH PART WITH THE ORIGINATING DRAWING NUMBER, INCLUDE IT ON THE PACKING LIST AND ON ALL INVOICES.

This is a corollary or extension to Axiom 4.

AXIOM 8

IF THE INFORMATION IS NOT IN YOUR HEAD
IT MUST BE RIGHT AT YOUR FINGERTIPS.

All of the information about the job must be immediately retrievable. You do
not have time to look for things, on the site or on the computer.

AXIOM 9

THE LONGER THE LEAD TIME ON CAPITAL
EQUIPMENT THE GREATER THE POSSIBILITY OF
UNFORSEEN DELAY, CHANGES OR PROBLEMS.

This goes for just about anything that has a probably of occurrence over a
period of time or events, such an earthquake or car accident.

CORRALARY TO AXIOM 9

A SUITABLE AMOUNT OF SLACK SHOULD BE BUILT INTO THE
SCHEDULE TO CONTROL WHAT GOES ON WITH AXIOM 9.

And more slack should be allowed the more complex, new or unique the
process.

AXIOM 10

SIZE MATTERS, BUT BIGGER IS NOT ALWAYS BETTER

AXIOM 11

THE CONTRACTOR SHALL VERIFY ALL DIMENSIONS,
AND WE MEAN EACH AND EVERY LAST ONE OF THEM.

This should go without saying, see also axiom 12.

AXIOM 12

MEASURE TWICE AND CUT ONCE

Timeless wisdom for anyone who has ever drilled a hole or cut a board. See also axiom 11.

AXIOM 13

BEWARE OF GREEKS BEARING GIFTS AND THE CONTRACTOR THAT TELLS YOU HE CAN MAKE UP FOR LOST TIME BY ADDING A SECOND SHIFT.

The most precious and unreplaceable resource is time. Do not start out by wasting it.

AXIOM 14

A GANG OF JOURNEYMEN GETTING OVERTIME DOES NOT ENTITLE THE SUBCONTRACTOR TO ADDITIONAL PROFIT, NOR DOES IT TAKE MORE OVERHEAD TO PROCESS A 10 HOUR TIMESHEET THAN ONE WITH 8 HOURS ON THE SAME SHEET OF PAPER.

Beware the contractor that thinks overtime should be more profitable than straight time. Over time is a penalty, due only to the worker for keeping him from the rest of his or her life.

AXIOM 15

THERE IS NOTHING ON A JOB THAT TAKES LESS THAN 4 HOURS OR COSTS LESS THAN $500 IN 1995 DOLLARS, ADJUSTED FOR INFLATION.

SO GET USED TO IT.

AXIOM 16

THE TAIL SHALL NOT WAG THE DOG

Beware the shifts and tides in the flow of power as the job progresses, the natural flow is from the owner (pre-bid) to the contractor (post-bid) and back to the owner (closeout)

AXIOM 17

IF YOU HAVE NOT UPDATED YOUR EV CURVES IN THE LAST THREE DAYS, YOU ARE LOSING CONTROL OF YOUR PROJECT.

A GLOSSARY TO THE ACRONYMS:

AASHTO	American Association of State Highway Transportation Officials
AC	Alternating Current
ACI	American Concrete Institute
AKA	Also Known As
ANSI	American National Standards Institute
API	American Petroleum Institute
ASAP	As Soon As Possible
ASME	American Society of Mechanical Engineers
ASCE	American Society of Civil (and/or) Chemical Engineers
AIA	American Institute of Architects
ASHRAE	American Society of Heating, Refrigeration and Air-Conditioning Engineers
ASTM	Formerly known as the American Society for Testing and Materials, now known as ASTM International
BOM	Bill of Materials
BS	Bachelor of Science
CA	Compressed Air
CD	Compact Disc
C&D	Construction and Demolition (used as a waste classification)
CM	Construction Manager
CNC	Computer Numerical Control
CP	Critical Path
CPM	Critical Path Method
CO	` Change Order
COB	Close of Business (end of a business day)
COR	Change Order Request
CYA	Cover Your Ass
CW	City Water
DC	Direct Current
DFR	Daily Force Report
DOD	Department of Defense
DOE	Department of Energy
DS	Drive Station

EMR	Experience Modification Rate
EMT	Electrical Metallic Tubing
EV	Earned Value
EVMS	Earned Value Management System
FAA	Federal Aviation Administration
FAR	Federal Acquisition Regulations
FCU	Fan Cooling Unit
FO	Field
FOGMA	Fuel Oil Gas and Maintenance (used as a construction cost category)
GC	General Contractor
GIGO	Garbage in = Garbage out
GPS	Global Positioning System
GL	General Liability
GM	General Motors
GPM	Gallons per Minute
HP	Horsepower
HR	Human Resources
HVAC	Heating, Ventilation, Air Conditioning
HTTP	Hypertext Transfer Protocol
HX	Heat Exchanger
IAW	In Accordance With
ID	Inside Diameter
IFB	Invitation for Bid
IFC	Issued for Construction
I/O	Input / Output
IRS	Internal Revenue Service
ISO	International Standards Organization
KW	Kilowatt
LEM	Lunar Excursion Module
LLC	Limited Liability Corporation
LOL	Laugh Out Loud
LOTO	Lock Out Tag Out
MCFD	Thousand (M) Cubic Feet Per Day (as a measure of gas flow)
MIG	Metal Inert Gas
MPL	Manpower Loaded (schedule)
MS DOS	Microsoft Disc Operating System
MSDS	Material Safety Data Sheet

MSL	Mean Sea Level
NASA	National Aeronautics and Space Administration
NASCAR	National Association for Stock Car Auto Racing
NEC	National Electrical Code
NEMA	National Electrical Manufacturers Association
NFPA	National Fire Protection Association
NOAA	National Oceanic and Atmospheric Administration
NPA	Notice of Potential Award
NPT	National Pipe Thread
NPTF	NPT (Female)
NPTM	NPT (Male)
NPV	Net Present Value
NTP	Notice to Proceed
OH&P	Overhead and Profit
OSHA	Occupational Safety and Health Administration
OT	Overtime
PC	Personal Computer
PDF	Portable Data Format
P&ID	Process and Instrumentation Drawing (or) Diagram
PLC	Process Logic Controller
PE	Professional Engineer
PPE	Personal Protective Equipment
PM	Project Manager
PO	Purchase Order
POC	Point of Contact
PV	Planned Value
QA	Quality Assurance
QC	Quality Control
R&D	Research and Development
RFI	Request for Information
RFID	Radio Frequency Identification
RFQ	Request for Quotation
RMS	Root Mean Square
ROM	Rough Order of Magnitude
SAE	Society of Automotive Engineers
SK	Sketch
SMACNA	Sheet Metal and Air Conditioning Contractor's National Association

SOV	Schedule of Values
SPE	Society of Petroleum Engineers
STI	Steel Tank Institute SCADA
TIG	Tungsten Inert Gas
T&M	Time and Materials
UL	Underwriters Laboratory
UMI	Unidentified Minor Items
URL	Uniform Resource Locator
VAC	Volts Alternating Current
VDC	Volts Direct Current
VCR	Video Cassette Recorder
WEGA	A Sony television product loosely named after the star Vega.
WTF	What the Fuck?
WWTP	Waste Water Treatment Plant

A GLOSSARY OF THE LATIN AND FRENCH PHRASES AND WORDS

Latin:

ad nauseum	(Repeat) until sick or nauseated
cur	Why?
quanto	With how much?
quando	When?
quem ad modum	In what way?
quibus adminiculis	By what means?
quid	What?
quis	Who?
res ipsa loquitir	It speaks for itself – in the roman senate, when evidence was so clear and needed no further comment. Actions speak louder than words.
status quo	The way things are, a newcomer cannot but upset the status quo.
ubi	Where?
veritas	Truth

French:

force majure	A major force, in contracts, a disruptive force beyond the control of either party, natural: earthquakes, fires, floods, or commercial: strikes, lockouts, supply disruptions
joie de virve	Joy of or exhilaration of life.
savoir faire	Expertise with, of special knowledge of.

BIBLIOGRAPHY

Ambrose, Stephen E, *D-Day June 6, 1944 The Climactic Battle or World War II,* New York: Touchstone, 1995.

Ambrose, Stephen, E, *Nothing Like It In The World, The Men Who Built The Transcontinental Railroad 1863-1869,* New York: Touchstone, 1995.

Bain, David Howard, *Empire Express, Building The First Transcontinental Railroad,* New York: Penguin, 1999.

Cicero, Marcus Tullius, *Selected Works, On Friendship, Translated by Michael Grant.* London: Penguin Classics, 1960.

Hanna, Awad S. et. al *Shift Work Impact on Labor Productivity,* California, University of Berkely, 1968.

Kidder, Tracy, *The Soul of a New Machine,* Boston: Little Brown and Company, 1981

McCollough, David, *The Great Bridge, The Epic Story of the Building of the Brooklyn Bridge,* New York: Touchstone, 1982.

McCollough, David, *The Path Between The Seas, The Creation of the Panama Canal 1870-1914,* New York: Touchstone, 1977.

Presidential Commission, *Report of the Presidential Commission on the Space Shuttle Clallenger Accident.* Washington DC: 1986

Salancik, Gerald R, and Jeffrey Pfeffer, *Who Gets Power-And How They Hold on to It: A Strategic-Contingency Model of Power.* Organizational Dynamics, Winter 1997.

Source Notes to the Charts, Tables and Images

Charts, tables and images presented herein which are not the original creation of the author have been pulled from sources within the public domain, or used with permission:

Robot throwing basketball, p. 21 used with permission, courtesy Carnegie Science Center, Pittsburgh PA.

Interconnect Diagram, P. 53 used with permission, courtesy Tetra-Pac, Inc.

LULL photograph, p. 220 used with permission, courtesy *JLG* industries, Inc.

Dante's Inferno, Illustration, p. 241 used with permission, courtesy Barry Moser.

INDEX

Printed in the United States
By Bookmasters